AF539502

Seed Quality

Seed Quality

Dr. Mahesh Kumar

RANDOM PUBLICATIONS
NEW DELHI (INDIA)

Seed Quality

ISBN 978-93-5111-362-1

Published in 2014 in India by

RANDOM PUBLICATIONS

4376-A/4B, Gali Murari Lal, Ansari Road
New Delhi-110 002
Phone : +91-11-43580356, +91-11-23289044
e-mail: randomexports@gmail.com, sales@randompublications.com, info@randompublications.com

Type Setting by : Keystoneprintads, Delhi-110051
Printed at Thomson Press (India) Ltd

Preface

In broad sense, seed is a material which is used for planting or regeneration purpose. However scientifically, Seed is a fertilized matured ovule together covered with seed coat is called seed or it is a propagating material i.e., part of agriculture, sericulture, silviculture and horticultural plants used for sowing or planting purpose.

Seed quality is determined by many factors, principally seed purity and germination. However, many other factors, such as the variety, presence of seed-borne disease, vigor of the seed, and seed size are important when considering seed purchase.

Seed purity is determined by the amount of unwanted material present in the pure seed. Contaminants such as noxious weed seed, unwanted crop seed or inert matter not only increase production costs, but also substantially reduce the quality and quantity of the harvest. If you purchase seed that has not been properly conditioned to remove unwanted weed seed, include in your decision-making the increased herbicide cost to control newly introduced noxious or common weeds.

Seed bags that contain the phrase "quick green" or "fast grass" often contain annual ryegrass. Annual ryegrass is just that - an annual. Annual ryegrass germinates rapidly and will grow throughout the summer but in almost all instances will die in the winter and leave the area bare the following year. In some cases, annual ryegrass may survive for a second year, but normally this is not the case. The normal use of annual ryegrass is as a cover or nurse crop for the desired turfgrass species.

The ability of seed to tolerate moisture loss allows the seed to maintain the viability in dry state. Storage starts in the mother plant itself when it attains physiological maturity. After harvesting the seeds are either stored in ware houses or in transit or in retail shops. During the old age days , the farmers were used farm saved seeds, in little quantity, but introduction of high yielding varieties and hybrids and modernization of agriculture necessitated the development of storage techniques to preserve the seeds. All in all, this is a complete guide explored the dissimilar facets of seed quality. The readers with extremely insightful and key information, it is hoped that this book proves to be highly resourceful.

I thank all members of my team who have helped in the preparation of the book. My special thanks go to "Random Publications" who have published the book.

– Dr. Mahesh Kumar

Contents

1

Introduction

Quick seedling emergence and even stands are essential to maximizing the yield of all crops. The use of high-quality, disease-free seed is the first step to producing good stands. The importance of using high-quality seed continues to increase due to changing production methods such as early planting, narrow rows, and no-till or reduced-tillage planting systems. The use of high-quality seed better ensures seedlings which emerge rapidly, tolerate adverse weather conditions, and resist disease.

SEED QUALITY AND SEED TREATMENTS

A number of factors affect the quality of seed. Seed-borne fungi, insect feeding damage, poor seed storage (temperature and humidity too high), and damaged seed coats can reduce the quality of seed. Seed quality may be reduced due to infection by fungi when harvest of the seed crop is delayed several weeks or when seed diseases occur due to wet conditions. Infected seed can be detected at harvest as discoloured, shriveled, or moldy in appearance.

Insect feeding that may occur in the field prior to harvest or in storage may affect seed quality. Germination of seed lots may be reduced by the insect partially eating the seed or opening the seed coat to fungal infections occurring at a later date. The seed coat is the primary protective layer surrounding the seed. Seed coat damage can result from rough handling during harvest, and cleaning or planting operations. Small cracks in the seed coat increase the chances of seed-rot by permitting water-soluble nutrients to escape into the soil which activate soilborne fungi that quickly enter the damaged seed.

Grain intended for seed should be handled as little as possible, dried using low temperature drying, and stored at moisture levels below 13 per cent. High humidity and temperature during storage promotes growth of fungi and insect damage. Seed lots should be closely monitored for mold growth and insect activity during storage. Seed treatments can play an important role in achieving uniform seedling emergence under certain conditions. The selective use of seed treatments can protect seeds or seedlings from early season disease and insect pests affecting crop emergence and growth. A few seed treatments may also be used to enhance crop performance during the growing season. Treatments containing Rhizobium inoculant have commonly been used to enhance the nitrogen fixing capability of legume crops. Increased germination can be observed when seed treatments containing fungicides are used when poor germination results from a fungal infection. Fungicide are available to protect seed and seedlings from many seed-borne or soil-borne pathogens. Seed treatments are most beneficial for seeds infected prior to planting or when cool, wet soil conditions exist at planting which results in delayed germination.

Seed treatments containing insecticides can serve to protect seed and seedlings from certain insect pests found in the soil. Cool, wet, seed beds resulting in delayed germination or plowing down green manure crops may warrant including an insecticide seed treatment. Seed treatments should not be considered a cure-all for the selection of poor seed lots. Seed treatments will not increase poor germination due to excessive mechanical damage, poor storage conditions, genetic differences in variety, or other damage.

The effect of a seed treatment on plant stands will depend upon seed quality as well as field conditions at planting. Extensive research has shown seed treatments can increase stands from poor quality seed or when planting conditions are less than optimum. However, seed treatments do not always increase yield. Generally, seed treatments result in increased yield by preventing death of seedlings causing frequent skips within crop rows. Grain producers should realise that any one seed treatment will not control all diseases or insects that may attack seeds and seedlings. Protection is also short lived, generally only lasting as long as it takes for the crop to emerge.

Before selecting a seed treatment it is essential for the grower to consider potential problems associated with the seed and the history of problems associated with each field. Each planting situation is different requiring consideration of seed quality; planting rate; tillage and seed bed preparation; soil moisture and temperature; time of planting; and the likelihood of rapid emergence.

GERMINATION TESTING AND SEED QUALITY

The purpose of laboratory testing of seed germination is to assess seed quality or viability and to predict performance of the seed and seedling in the field. Seed processed for sale must be tested by a qualified laboratory under the Association of Official Seed Analysts Rules for testing seeds. Several different kinds of testing are available depending on the type of seed to be tested, the conditions of the test, and the potential uses of the seed. The most common tests are the warm germination test, cold germination test, accelerated aging test, and the tetrazolium test. Each test is designed to evaluate various qualities of the seed. Factors that can affect the performance of seed in germination tests include; diseased seed, old seed, mechanically damaged seed, seed stored under high moisture, and excessive heating of seed during storage or drying. In most cases a seed treatment will improve germination of seed only if the poor quality is due to seed-borne disease. The most common test is a warm germination test because it is required by seed laws to appear on the label. The percentage of germinating seed in a warm germination test must be printed on the label of the seed if it is to be sold as seed. Germination is defined as: "the emergence and development from the seed embryo of those essential structures which are indicative of the ability to produce a normal plant under favorable conditions."

The warm germination test reflects the stand producing potential of a seed lot under ideal planting conditions. Usually 400 seed from each seed lot are placed under moist conditions on blotters, rolled towels, or sand and maintained at 77 degrees F for about seven days. At the end of this period the seedlings are categorized as normal, abnormal, or diseased, and dead or hard seeds. The percentage germination is calculated from the number of normal seedlings from the total number of seeds evaluated. The cold germination test is designed to measure the ability of seeds to germinate under high soil-moisture content and low soil temperature. This vigour test simulates early season adverse field conditions and usually represents the lowest germination that would be expected from a seed lot planted under such conditions.

Actual field germination would normally fall between the cold test result and that of the warm germination test. Seeds are planted in a sand-soil mix at high moisture content and maintained at 50 degrees F for seven days. The test is then placed at 77 degrees F for four days. The percentage of healthy seedlings that emerge at least one inch above the soil is reported. The accelerated aging test estimates the carryover potential of a seed lot in warehouse storage. The seeds are exposed to high temperatures and high relative humidity for short periods of time that cause seed deterioration. Seed are suspended over water in a chamber for 72 hours (wheat and soybeans) or 96 hours (corn) then tested in a standard warm germination test. This test only would be used on seed whose longevity was in question.

The tetrazolium test is a "quick test" for seed lot viability. It is useful when an approximate germination percentage is needed immediately. Seeds are soaked overnight in water then treated with tetrazolium to give an indication of viable, abnormal, and dead seeds in the seed lot. This test will not detect seed-borne disease, thus is limited in its ability to estimate seed quality. This test is highly reliable for determining viable seed of corn, wheat, oats, barley, and other grasses.

If growers wish to use bin-run seed and has not had a germination test conducted by a competent laboratory, they can get an approximate germination test using the following procedure in their own home.

Place two paper towels in the bottom of a dish or tray, one on top of the other; wet the towels thoroughly and tilt the tray up on one end so that excess water runs off the tray. Select a random sample of 100 seeds from the seed lot and place them in between the moist paper towels. Put the tray in a plastic bag and tie the end shut to prevent the towels from drying out.

Place the tray in a location of diffuse, not direct light, such as a north window. The location should be warm enough for good plant growth. An ideal location is with well growing house plants. After five to seven days, open the plastic bag and count the number of germinated seed with intact tap roots and shoots. Do not count moldy seed or diseased seedlings. Testing 400 seeds in this way will give a good indication of the germination percentage.

Choose the Right Fungicide for the Problem

The basic component of most seed-treatment products is a fungicide. Insecticides are contained in some products to give insect protection when such control is desirable. The grower should select a seed treatment product that will control the most likely pest to be encountered within a given field. It is important to note that not all fungi are alike. One fungicide seed treatment may be very effective against one fungus but will have no effect on another. Rates of active ingredient used also affect a seed treatment performance on different disease problems. Fungicides have fungicidal activity, that actually kills the fungus, or fungistatic activity, that slows the growth of the fungus to prevent extensive damage. Fungicides used in seed treatment are of three types: protectants that generally prevent infection of seeds and seedlings by soilborne fungi, surface disinfestants that kill fungi and fungal spores on the surface of the seed coat (seed-borne fungi) and "systemic" disinfectants that kill fungi already established within the seed (seed-borne fungi).

Insecticides used in seed treatments act as surface protectants killing insects as they feed on the seed or emerging seedling. The marketing of seed treatments by manufacturers may result in the same chemical being sold under several brand names and by several different companies. All chemicals have three names on the packaging; chemical name, common name, and trade or brand name. The chemical name represents the molecular structure of the active ingredient and is often long and complex.

The common name is used by the manufacturer in lieu of using the longer chemical name. The trade or brand name is used by the manufacturer in product marketing. The chemical name (R,S)-2-[(2,6-dimethlyphenly)-methoxyacety-lamino]-proprionic acid methyl ester has the common name mefenoxam and is marketed under the trade or brand name Apron XL LS. Using mixtures of different seed treatment fungicides or insecticides is a common practice. Mixing two or more fungicides or insecticides results in a product that will control a broader spectrum of seed pests.

Applying Seed Treatments

Seed treatments come in a variety of formulations: dry flowables (DF), liquid flowables (LF), true liquids (TL), emulsifiable concentrates (EC), dusts

(D), wettable powders (WP) and others. These materials may be used in slurry and mist type seed treaters upon mixing with water as specified in instructions on the product label. Special adhesives are incorporated into the dust formulations to adhere pesticide particles to the seed surface.

Commercial Treatment

Commercial seed treatment is often desirable due to the specialized equipment required to properly apply treatments or to treat large volumes of seed. The primary concern of the commercial treater is equipment calibration to ensure delivering the proper amount of active ingredient on the seed. This has become especially important with more modern fungicides that require only very small amounts of material (less than 0.5 oz) per hundred weight of seed. Detailed information on the process of treater equipment calibration can be obtained from the equipment manufacturer. Seed treatment manufacturer labels provide information related to mixing formulations and mixing procedures.

Information needed to properly calibrate a seed treater includes:

- The correct labeled rate of the seed treatment material that will be used,
- How many pounds of seed is dumped each time the weigh pan arm dumps,
- The size of the cup being used in the treater.

This information will allow you to calculate the correct application rate for the seed and treatment material used. A new generation of seed treater is becoming available that offers precision distribution of treatment chemicals on individual seed through atomization. These advanced controlled droplet application (CDA) machines offer complete seed coverage with lower water volume, shorter drying time, and continuous seed flow. These machines also reduce chemical exposure to employees and are capable of applying current and future products that may include both chemicals and biologicals.

On-farm Application

Many seed treatment materials are available for on-farm use. These are known as hopper-box or planter-box treatments. Most hopper-box treatments are dry treatments formulated with talc or graphite which adheres the treatment chemical to the seed. Liquid hopper-box treatments are available as a fast-drying formulation. Good seed coverage is required for maximum benefit from any seed treatment formulations.

Obtaining thorough seed coverage can be difficult when attempting to treat seed in the field at planting time. Most seed treatment companies recommend that the planter box be filled only half full with seed, and half the required amount of treatment product added. Mix the product into the seed volume with a stick. Then add the remainder of seed to the planter box and add the rest of the required amount of product, again mixing thoroughly. Good results are also obtained by premixing seed with the product in a suitable container, then pouring treated seed into the planter box.

On-the-farm treaters are available that apply liquid or dry formulations to seed as it passes through an auger from the transport bin or truck to the planter boxes. These treaters are a very convenient way to apply seed treatment onto bulk seed. For additional information on on-the-farm treaters, contact a representative of the company selling planter-box treatments.

Labeling Treated Seed

Once seed is treated it can no longer be used for food feed or other purposes. Under the federal and state seed laws, the following information is required to be shown on the label: a word or statement in type no smaller than eight points indicating that the seeds have been treated; the commonly accepted, chemical (generic) name of the applied substance including the application rate, and a caution statement if the substance used in such a treatment has an amount remaining with the seed that is harmful to humans or other vertebrate animals. Seed treated with highly-toxic substances shall be labeled in red ink to show a statement such as "Poison Treated." In addition, the label should show a representation of skull and crossbones. Seed treated with less toxic substances shall be labeled to show a caution statement in no smaller than eight point type such as "Do not use for food, feed, or oil."

Colouring Seed Treated with Poisonous Substances

All treated seed must be coloured an unnatural colour to distinguish it from untreated grain and prevent inadvertent use as food for man, feed for animals, or for oil purposes. All treated seed must be coloured with a dye, colour coat, or colour film which must cause no apparent injury to seed germination or danger to personnel processing or using the seed.

Dyes are designed to stain the seed with colour. Colour coat pigments cover the seed with colour, and colour films are made from a polymer that creates a coloured film around the seed. Seed treatment companies currently offer a wide variety of colour and surface texture options to seed processors.

SAFETY PRECAUTIONS

Seed treatment chemicals are pesticides and should be handled as such. Product labels provide information on safe handling and application. Be sure to follow all instructions as listed.

Hazards Associated with Handling

Read the label and follow instructions carefully. Over treatment may injure the seed and under treatment will not control the pests in question. The label will also provide information on the toxicity of the active ingredients in the form of code words. The label will provide basic information on first aid, environmental hazards, directions for use, proper storage and disposal of containers, and pesticide and general product information. Use the label and be familiar with it. Material safety data sheets (MSDS) are available from the company for each of the products sold. Consult the MSDS for more detailed information before applying the material.

When handling toxic substances, operating personnel must be provided with protective clothing such as coveralls, cap, protective glasses, rubber apron, rubber boots, rubber gloves, and a respirator designed for use with the substance. Personnel should not inhale the dust or vapour nor permit the material to contact the skin or eyes. The operator should wash thoroughly with soap and water before eating and smoking. Bathe immediately after work and change all clothing and wash clothing thoroughly with soap and hot water before reuse. In case of contact, immediately remove contaminated clothing and wash skin thoroughly with soap and water. A safety shower should be installed in the immediate vicinity of the treater.

An exhaust system should be installed to remove vapours and dust from the operating area. The exhaust air should discharge into a cyclone or bag-type dust collector. The treater should also be vented and tied into the exhaust system. Seed treaters should be isolated and not operated in the vicinity of other personnel or farm commodities that are to be used for food, feed, or oil. Special multiwall paper bags or tightly woven bags are recommended for seed that has been treated with toxic substances. Seed should be thoroughly dry before going into the bag, as excessive moisture can cause rapid deterioration of the seed. Store treated seed in a cool, dry place away from food or feed products. Seed treaters should be thoroughly cleaned after use as some of the pesticides are corrosive and others settle out and cause clogging of the equipment. Do not run contaminated water into a stream or public sewer, but discharge into a shallow ground pit.

Hazards Associated with the use of Treated seed

The user of treated seed should read and follow the label carefully. The planter hoppers should be filled outdoors. Do not breath dust or fumes from treated seed or get in eyes or on skin. Wash thoroughly with soap and water after handling treated seed. Use treated seed as seed and do not introduce them into the food or feed channels because serious injury to livestock and humans may occur. Follow instructions on the label for disposal of seed treatment containers and bags containing treated seed.

Do not reuse pesticide containers. Triple rinse and, if possible, recycle containers. Containers can be punctured and disposed of in a sanitary landfill or incineration, or, if allowed by state and local authorities, by burning. If burning, stay out of the smoke. Old seed with low germination should be destroyed by burying at least 18 inches deep in an isolated area away from the water supplies. Treated seed exposed on soil surfaces will be hazardous to birds and other wildlife.

Biological Seed Treatments

Recent research has focused on biological control of certain seed and seedling diseases of crops. The most promising biological control agents tested have been either fungi or bacteria. These biological seed treatments control seed pests by parasitizing the pest organisms, competing for food on the root system, or producing toxic compounds that inhibit pathogen growth. Several biological seed treatments are currently on the market.

Insecticide Seed Treatments

A limited number of insecticides are labeled for use as seed treatments to protect seedlings of various field crops from soil inhabiting insects.

SOIL INSECT PESTS

Soil insect pests are generally more of a problem on corn and soybeans than on small grains and forages. Soil-applied insecticides recommended for control of rootworm larvae also may prevent early stand losses due to wireworms, grubs, seedcorn maggot, and seedcorn beetles. If a soil-applied insecticide is used at planting, then a seed applied insecticide may not be necessary. The label of the soil-applied insecticide to determine which insects are controlled before purchasing a seed treatment insecticide. Stand losses to these soil insects is generally not significant enough to warrant routine use of soil-applied insecticides in first-year corn. The use of insecticide for first-year corn may change in the future due to a biotype of the western corn rootworm beetle. Benefits from suppression of early-season soil pests may be significant in:

- No-till or reduced-tillage conditions with poor weed control where a high risk exists for a given insect problem such as wireworms and grubs in corn following established sod,
- Continuous corn where early-season pests may cause unacceptable stand losses.

Seedcorn Maggot

Seedcorn maggot or the bean seed maggot are small, yellowish-white, legless fly larvae found feeding on corn seeds. Extensive feeding by these maggots will cause a reduction in stand. In general, seedcorn maggot problems are most likely to occur in situations where:

- High organic matter or decaying vegetation attracts egg-laying adult flies.
- Cool, damp soil conditions delay seed germination and prolong the period vulnerable to maggot attack.

No-till production systems have not been found to increase the likelihood of problems with seed maggots. Several products are formulated for application prior to planting seed on the farm as planter-box or hopper-box treatments.

Seedcorn Beetle

Fig. Seedcorn beetle

Partially eaten seeds, loss of germination, or stunted seedlings in the presence of small (1/4 to 1/3 inch long) brown ground beetles indicate a seedcorn beetle problem. As with seedcorn maggots, damage is most likely to occur under cool, damp conditions where seed germination and seedling development are delayed.

Wireworms

Wireworms are the larval stage of the "click beetles." The term "wireworm" applies to a complex of species with life cycles requiring one or more years per generation. Wireworm populations affecting corn are most severe in fields following sod or fields having a prolonged grassy weed problem. The larvae pass through a number of life stages, or instars. The earliest stages are very small and white; the latter stages have a characteristic hard-shell appearance and are yellow-brown in colour.

Full-grown larvae range from 1/2 to 1 inch long depending on the species. Wireworm's damage corn by feeding on germinating seeds and young seedlings and may bore into stalks at the soil level. Stand loss may be significant in fields with high populations. In fields where a soil-applied insecticide is not used, especially first year corn fields, application of seed treatment insecticide is strongly recommended.

WHEN TO TREAT SEED

Corn

Treatment is recommended for all seed corn to prevent or reduce seed decay and seedling blights. The germination and early growth of the seedling is a critical period in the life cycle of corn. Soilborne organisms may invade and kill the embryo before germination, or when invasion occurs later, the seedlings may be destroyed before or after emergence. Seedlings surviving fungal infections at germination are often less vigorous.

Corn seedling diseases are more prevalent in cold, wet soil than when soil temperatures are above 55 degrees F. Therefore, early planted corn needs the added protection of a good seed treatment fungicide. Seedling blights tend to be more severe in fields with poor seed bed preparation such as in no-tillage or reduced-tillage fields. Corn seed is generally treated with a fungicide or fungicide-insecticide combination by the seed producer or seed processor.

Corn smut, leaf blights, stalk and ear rots, and virus diseases are not controlled by seed treatment fungicides.

Small Grains

Seed treatment of small grains is recommended to control smut diseases and to reduce seed decay and seedling blights. The smut diseases include stinking smut (common bunt) of wheat, loose smut of wheat and spelt, and the loose and covered smuts of oats and barley.

All of the wheat varieties planted are susceptible to stinking smut (common bunt). Wheat and oat varieties have resistance to some, but not all, races of loose smut (wheat) and loose and covered smuts (oats). The only sure method of smut control is by using an effective seed treatment fungicide. Spelt is extremely susceptible to loose smut and all seed should be treated with an effective fungicide.

Seedling blight phases of head scab (*Gibberella zeae*) and glume blotch (*Stagonospora nodorum*) contribute to poor stands of small grains. Both seed infecting fungi contribute to poor quality seed by causing lightweight, shriveled grain. Since these seed-borne diseases are common, all small grain seeds should be treated to control seedling blight when infected seed are planted.

Soybean

Seed treatment for soybean is recommended in three disease situations:

- When poor quality seed is used for planting,
- When early season control of Phytophthora damping-off is needed,
- When planting early into cool, wet soils, especially in reduced tillage.

The major cause of poor-quality soybean seed is Phomopsis seed rot. Infected seed may be visibly moldy, yet others appear healthy. Purchasing certified seed with the per cent germination, the variety name, and the seed treatment material used listed on the label is one way to ensure quality seed is planted. Germination percentage of bin-run beans is generally unknown unless they are tested.

Seed lots with greater than 80 per cent germination generally do not benefit from seed treatment when planted at recommended seeding rates in warm soil. Those with less than 80 per cent germination may benefit from seed treatments by increased stand and, possibly, increased yield. Decisions to use poor-quality seed should be based on the fact that any seed treatment fungicide effective against seed-borne diseases will not increase germination more than 20 per cent.

If bin-run beans of 50 per cent germination are treated with a fungicide, the grower should not expect over 70 per cent germination in the field. Planting rates can be increased to compensate for seed of low germination.

A number of fungicides are effective in controlling Phomopsis seed rot. These materials have varying degrees of activity against seed and seedling diseases.

Phytophthora damping-off is a serious disease of soybean seedlings in the more heavy, poorly-drained soils. The increase in no-tillage and reduced tillage has increased the incidence of Phytophthora damping off in the state. No-till soils remain wetter longer in the spring and less precipitation is required to saturate them compared with plowed soil during the early stages of crop development.

Extra moisture in the soil pore spaces favors germination of *Phytophthora*. It is recommended that seed treatments be used even when planting *Phytophthora* resistant varieties to ensure good stand establishment. Rhizoctonia seedling blight has caused considerable stand losses over the last decade. Damage from *Rhizoctonia* appears to be a stress related in that factors that adversely affect germination and establishment of seedlings predisposes the plants to infection. Stress from planting in dry soil, herbicide injury reduced tillage fields and shallow planting appears to increase the potential for damage by *Rhizoctonia*. Adequate seed bed preparation, soil fertility, and soil moisture for rapid emergence lessens the potential for problems from *Rhizoctonia*.

Sclerotinia stem rot may be introduced into new fields by sclerotia-contaminated seed lots or infested seed. Seed should be well cleaned to remove sclerotia and soybean seed should be treated with a fungicide to eradicate the fungus on infested seed.

Alfalfa and other Small-seeded Forage Legumes

Although a number of fungicides are registered for use as seed treatments of alfalfa, clover, etc., these crops usually do not respond to seed treatments under field conditions. Some increases in stand may result but these increases have not been reflected in increased yield. The one notable exception to this is the use of fungicide seed treatment on alfalfa for control of seedling damping-off caused by *Phytophthora*.

Significant improvement in stand establishment of alfalfa has been noted in research plots using treated seed. In areas where Phytophthora root rot has caused a serious problem in reducing alfalfa stands or preventing establishment of alfalfa, a variety with resistance to *Phytophthora* that has been treated with a *Phytophthora* specific seed treatment is highly recommended.

Rhizobium Inoculation and Seed Treatments

Rhizobium and Bradyrhizobium are bacteria which can fix nitrogen from the air or collect nitrogen from the soil water solution and incorporate it into compounds that plants can use. The nodules that form on soybean roots are the sites where the bacteria reside. Rhizobia are present in soils, but the commercial strains and some newer strains identified by the USDA are much more efficient at fixing nitrogen. The bacteria may be applied to the seed or placed in the seed furrow at planting. Insecticide and fungicide seed treatments can have a negative impact on inoculants applied to soybean seed. Several factors affect the impact of seed treatments on seed applied inoculants including: toxicity of material, toxicity of carrier or formulation, length of time inoculant is in contact with seed treatment, and formulation of inoculant. Insecticide seed treatments tend to be the most toxic of all seed treatments to inoculants. The fungicides themselves are not necessarily the problem, but the formulation or carriers may inhibit Rhizobial growth and colonization.

Using fungicides in combination with inoculants can be successful if some precautions are taken. If a specific material is highly toxic to inoculants, then placing the inoculant directly in the furrow and not on the seed is recommended. Treating the seed first and allowing the seed treatment material to dry followed by using humus preparation of inoculant may be successful.

Liquid inoculants and fungicides should not be mixed and applied simultaneously. Producers should minimize the time the inoculants are in contact with the seed treatment fungicides prior to planting.

Corn Diseases

Damping-off and seed decay. Germinating corn seed can be attacked by several seed-borne and/or soilborne fungi. Pre-emergence and post-emergence damping-off caused by fungi is most common in poorly drained, cold soils. Seed rots and seedling blights are commonly caused by *Pythium* and *Fusarium* species, and may be caused by *Penicillium* and *Bipolaris* species. All of these fungi can rot seed prior to germination. Infected seedlings typically show a marked softening of stem tissue at the soil line.

Kernels with surface cracks caused by mechanical harvesting are especially susceptible to seed rots caused by soilborne pathogens. Broad spectrum, protectant-type fungicides are highly recommended for both corn and sorghum, especially when early planting in cold, wet soils is attempted. Pythium seedling blight is most significant under these conditions.

Most seed corn companies treat their seed prior to bagging. However, planter-box formulations of these materials are available for use by the grower on untreated field, sweet, and popcorn seeds.

Soybean Diseases

Phytophthora damping-off and root rot is caused by the fungus *Phytophthora sojae,* and is particularly a problem in low, poorly drained, clay soils. However, the disease can be encountered on a variety of soil types if the soil remains wet for several days soon after planting.

Phytophthora can attack soybean plants at any stage of development and stands can be reduced by seed rotting and pre-emergence damping-off. Young plants can be killed soon after emergence. Brown, water-soaked stems and yellow, wilted leaves are the primary symptoms of post-emergence damping-off.

Phytophthora survives in the soil as thick-walled spore, called oospores. Early in the growing season, when the soil remains wet for several days, these spores will germinate, producing a second type of spore called a sporangium. The sporangium then germinates to produce a third type of spore called a zoospore in large numbers.

Zoospores are attracted to chemicals released by soybean roots and swim through a thin film of water until a root is encountered. The zoospores produce a hyphae (a threadlike structure) and infect the soybean root. The fungus grows in the roots and eventually into the plant stem. As the plant dies, oospores are formed by the fungus.

Pythium damping-off can be caused by several species of *Pythium*. Pre-emergence damping-off, seed decays, and root rots of soybean all can be caused by this group of fungi. Pythium rots can occur on soybeans at any stage of plant development but are much more prevalent on seedlings. Diseases caused by *Pythium* can occur over a wide range of temperatures, but are most common during cold periods with high soil moisture.

Seedlings infected with *Pythium* often fail to emerge. Stems of infected seedlings appear water-soaked and translucent above the soil line. Diseased areas later turn brown and appear shrunken; eventually, the stem and smaller roots decay and the seedlings die.

Pythium species survive also as oospores. During periods of high soil moisture, the resting spore germinates and infects seeds or young plants in much the same way as *Phytophthora*. Germinating soybean seeds release a wide variety nutrients or by-products that can stimulate the growth and attract spores of the fungus.

Many different seed treatment materials will effectively prevent losses from Pythium seed rot and seedling blights.

Rhizoctonia seedling blight is caused by the fungus *Rhizoctonia solani,* and may occur at any time, but a protected cool period of low soil moisture followed by warming soil temperatures with a brief rainy period favors disease development. Seedlings under stress are more susceptible to Rhizoctonia seedling blight. Example of stress are herbicide injury or desiccation during early seedling growth. Seedlings and young plants are most often affected.

Infected plant stems and older roots appear reddish brown and have sunken lesions. Stems may be girdled just above the soil line; tissue thus damaged may appear cracked or cankered. In dry, windy weather, severely infected plants wilt and die rapidly.

Rhizoctonia solani survives in the soil as sclerotia. The fungal mycelium body itself may also survive in soil or in association with old plant residue. Growth of the fungus in soil depends on soil nutrient supply, pH, moisture, and temperature.

As the fungus grows in soil, it can encounter and infect germinating soybean seeds or young plants. Several seed treatment materials can control the seed rot phase of this disease, however, none are highly effective against the post-emergence or stem rot phase of the disease.

Phomopsis seed rot is a common seed disease and is caused by the seed-borne fungi, *Phomopsis longicolla, Diaporthe phaseolorum* var. *sojae,* and *D. phaseolorum* var. *caulivora*. Infected seeds typically germinate poorly or not at all.

Fig. Phomopsis seed rot of soybean

Severely infected seeds appear shriveled, cracked, and elongated, and may be covered with a white, moldy growth. However, seeds may be infected and fail to show any symptoms at all. Infection of seeds is most common when warm, wet weather delays harvest. When moldy seeds are planted, seedling death results in poor stands.

Generally, higher soil temperatures favors development of seed rot and seedling blight. Seed treatments can increase the germination rate up to 20 per cent. Seed lots with less than 80 per cent germination should not be used.

Sclerotinia

Recent research demonstrates that Sclerotinia white mold can be introduced into new fields on infested soybean seed. In addition, this fungus, *Sclerotinia sclerotiorum*, forms hard, black, small, irregular-shaped sclerotia both inside and outside of infected soybean plants.

These sclerotia are harvested with the seed or returned to the soil with plant debris at harvest. Seed should be well cleaned to remove sclerotia and treated with appropriate fungicide seed treatment to eradicate the fungus from infested seed.

Wheat and Spelt Diseases

Common bunt or stinking smut is caused by the seed-borne fungus *Tilletia laevis*. The disease causes reduced wheat yields and grain quality. Common bunt was once the most prevalent disease of wheat, however, today the disease is rarely found due to the widespread use of effective seed-treatment fungicides.

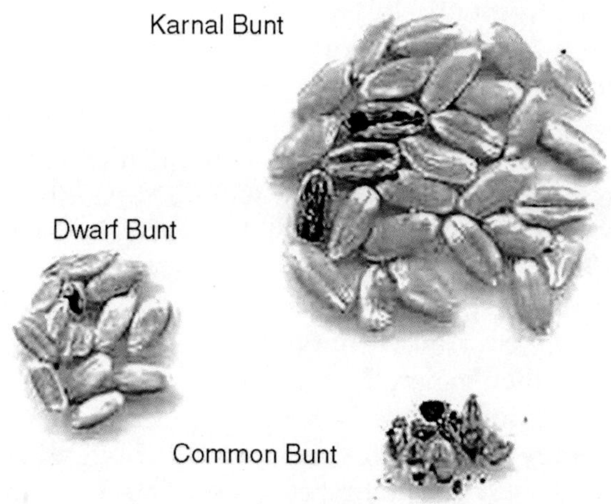

Fig. Bunt of wheat

Heads of diseased plants may appear stunted and are noticeably thinner and darker in colour than healthy, disease-free heads. "Bunt balls" (kernels containing millions of fungal spores) replace the kernels in the flowering heads and spread the glumes much farther apart than normal. Bunt balls approximate the size of normal kernels but tend to be more spherical and off-coloured (typically gray-coloured). When the bunt balls are ruptured at harvest, they release spores into the surrounding air and contaminate the surface of healthy seed.

Bunt fungi survive as resting spores on contaminated seed. When contaminated seeds are planted, bunt spores germinate in the presence of moisture and infect the wheat seedlings. Once established, the fungus grows systemically in the plant. At flowering, entire flowering heads may be infected, and the fungus converts the developing kernels into bunt balls. Healthy seeds are contaminated by the release of the spores at harvest.

Loose smut, caused by the fungus *Ustilago tritici*, has little effect on seed quality. However, it can result in substantial yield losses if left unmanaged. Symptoms are most noticeable between heading and maturity of the crop. Diseased flowering heads are conspicuously darker when compared to healthy, green heads. Also, diseased heads typically emerge earlier. Diseased spikelets may be entirely transformed into dark fungal spore masses.

As the head emerges, floral tissue is torn and the spores are released. Eventually only a spike is left where a normally healthy head should be. Spores of the fungus are dispersed by wind to nearby healthy flowers where they initiate new infections during wet weather. Unlike bunt, the loose smut fungus infects the developing kernels without causing noticeable damage.

Ustilago tritici survives in infected wheat seed. The fungus becomes active when the seed germinates and grows into the growing point of the developing wheat plant. Fungal growth closely follows the plants' growth. When the flowers form, the fungus again sporulates and starts the cycle again. Use a systemic fungicide seed treatment to eliminate this pathogen from infected seed.

Head scab, caused by *Fusarium* spp., is frequently a major cause of poor quality wheat seed. Infected seed appear whitish to reddish in colour and are nearly always shrunken and light in weight. Scabby kernels may be dead or the young seedling may be killed by seedling blight after emerging from the soil. Roots of plants killed by seedling blight appear light to reddish brown in colour and may be covered with mold.

If they survive, they generally lack vigour and may produce only a few weak tillers. Planting infected seed early when soil temperatures are above 60 degrees F generally increases loss from seedling blight. The best control of the seed-borne phase of scab is to first clean seed lots to remove all lightweight seed, thus increasing test weights; and second, treat with a fungicide that is effective in controlling seedling blights.

Fig. Head scab of wheat

Stagonospora glume blotch, caused by the fungus *Stagonospora nodorum* has increased in the US in recent years probably due to expanded use of nitrogen fertilizers, semidwarf wheat varieties, and reduced tillage.

Symptoms can develop throughout the growing season on all above-ground plant parts. Initial symptoms are small chlorotic flecks usually on lower leaves or those in contact with soil. The glumes of the wheat heads become infected in late spring before flowering. Lesions generally begin at the glume tips and are dark brown in colour. During favorable weather, the fungus penetrates the glumes and infects the seed causing severe shriveling of the grain.

Stagonospora survives on seed, straw, and/or volunteer wheat. With the onset of moist weather, spores are produced and are spread to healthy wheat plants by splashing rain. Glume blotch is most common in relatively warm weather when the temperature is between 70 degrees F and 80 degrees F.

Barley Diseases

True loose smut of barley is caused by the fungus *Ustilago tritici*. The disease cycle is exactly the same as that of loose smut of wheat. Infection occurs when the flowers develop. The seeds become infected internally and must be treated with a systemic fungicide to eliminate the pathogen.

Semiloose smut is caused by the fungus, *Ustilago avenae*. The disease typically causes small losses due to the use of resistant cultivars and fungicide seed treatment. It is most common when soil temperatures are cool and the soil is dry. Distinguishing semi-loose smut from true loose smut in the field is extremely difficult. Diseased heads typically appear earlier than healthy heads. Millions of fungal spores may be formed in diseased heads.

Ustilago avenae survives as teliospores on the seed surface. At seed germination, the fungus becomes active and infects young seedlings prior to their emergence from soil. As the plants develop, the fungus grows within the growing point of the young plant. At the boot stage, the fungus transforms the kernels and glumes into dark spore masses very similar to those produced by the true loose smut fungus. Spores are readily dispersed by wind or the action of harvesting equipment. Healthy seeds are contaminated upon contact with the spores.

Covered smut caused by the fungus, *Ustilago hordei*, can result in significant yield losses if left untreated. It is a relatively common disease throughout the barley-growing regions of the world. Problems with the disease occur over a wide range of temperatures and are most common under moist soil conditions.

Ustilago hordei survives as thick-walled resting spores on barley seed. Typically, spore germination coincides with seed germination, it is at this time that infection occurs. Once established, the fungus invades the actively growing tissue and keeps pace with growth of the plant. At flowering, the fungus grows through the floral tissue and forms masses of spores in place of healthy seed. Each kernel becomes a spore filled "smut ball" that is enclosed by a thin membrane until plant maturity. At harvest, spores are liberated and may contaminate healthy seed.

Barley stripe, caused by the seed-borne fungus, *Drechslera graminea*, is most common during periods of rain or high humidity. Few seeds are produced on infected plants, and losses generally coincide with disease levels in a particular field. Properly treated seed is the primary means of controlling this disease. Symptoms are first noticeable on the first two or three leaves produced by the plant and on many leaves produced later.

A yellow stripe on the leaf sheath and basal portion of the leaf blade is common. Later, the stripe may extend the entire length of the leaf and the leaf may die. At heading, diseased plants are often stunted and appear a light tan colour in contrast to the larger, green healthy plants. Spikes may be deformed or may fail to emerge. At heading, spores are produced on diseased leaves and spread by wind to nearby heads. Grain produced by diseased plants is typically brown and shriveled.

Drechslera graminea survives entirely as mycelium in the outer layers of the seed. When the seed germinates in moderately moist soil at temperatures below 55 degrees F, the fungus mycelium penetrates and infects the seedling. The fungus grows within plant tissue and occupies the actively growing plant parts.

Net blotch and spot blotch, common diseases of barley, are caused by the fungi, *Drechslera teres* and *Bipolaris sorokineana*, respectively. Net blotch is most common during rainy and humid weather, and early flag leaf infections can result in both reduced grain yield and weight.

A net-like pattern on sheath leaves or flag leaves is the most recognizable symptom of the disease. Initially, small spots or streaks appear on the leaves, followed by an expansion into conspicuous longitudinal and transverse streaks, eventually combining into a network. Brown areas surrounded by greenish-yellow margins are commonly observed on infected leaves. Eventually, entire leaves may wither and die.

Drechslera teres survives in infected seed or on crop residues. Infection of seedlings is most common at 34 degrees F to 60 degrees F. Once infection is established, the fungus produces spores on leaf tissue during periods of high relative humidity. Spores are released and carried by wind to nearby barley plants where new infections can result. Infection of leaves can occur when leaves are wet for periods of 5 to 15 hours within a 46 degrees F to 91 degrees F temperature range.

Spot blotch is most commonly a problem during warm, humid weather. Yield losses of up to 36 per cent have been attributed to this disease when the flag leaf has become severely diseased early in the season and has died prematurely. Seedling blight (*B. sorokineana*) in winter barley is favored by planting early in warm soil. Delayed planting of winter barley in cooler soil helps prevent blight.

Small brown spots surrounded by a yellow halo is the characteristic symptom of the disease. Both leaves and leaf sheaths on plants of all ages can become infected. Eventually, the spots coalesce and cover large areas of the leaves. Minute fungal fruiting structures containing spores can sometimes be observed in the larger spots. Severely diseased leaves may eventually die.

Bipolaris sorokineana survives in seed or infested crop residue. Leaf infection results from airborne spores produced on the seed or on residue. Epidemics of spot blotch can occur when there are prolonged wetness periods (greater than 16 hours) and the temperature is above 70 degrees F. Note that net blotch is more common in cooler, humid weather, as opposed to spot blotch which prefers warm, humid weather. Most common broad-spectrum seed treatment fungicides are effective in eliminating seed-borne *B. sorokineana* and *D. teres* as well as preventing losses from seedling blight caused by them.

Oat Diseases

Loose smut and covered smut, caused by *Ustilago avenae* and *U. segetum*, respectively. When disease outbreaks are severe, both yield and grain quality can be significantly reduced. However, fungicide seed treatments effectively control both diseases.

Symptoms of these diseases are very similar. Smutted oat panicles emerge from sheaths as olive-brown to black fungal spore masses. Smutted panicles remain more compact than healthy ones, and usually all spikelets on a diseased plant are affected. Smutted plants are often stunted and easily detected in a recently headed field of oats. The thin membrane enclosing the spore masses of *U. avenae* usually rupture and disintegrate after panicle emergence as opposed to those of *U. segetum*, which can persist longer.

Alfalfa Diseases

Phytophthora damping-off and root rot is caused by the fungus, *Phytophthora megasperma f.sp. medicaginis*. Standing water, poor drainage, and generally wet soil conditions favour disease development. This disease can be lethal to seedlings and established plants.

Seedlings may fail to emerge or can be killed after emergence. A conspicuous yellowing of leaves, particularly lower ones, is a characteristic symptom. Infected seedlings often wilt. Seedling tap roots have dark brown to black lesions and may be rotted 2-4 inches below the crown. A yellow discolouration of internal root tissue is often present. Tap roots may appear discoloured. The life cycle of *Phytophthora* on alfalfa is similar to that on soybean, however, the fungi are completely different and do not attack the other plant host. Seed treatments are labeled for control of Phytophthora damping off. However, to protect stands beyond the seedling stage, alfalfa varieties with resistance to the fungus should be used.

SEED DESCRIPTION AND VARIATION

DESCRIPTION AND VARIATION

A perennial herb that ranges from 2–6 feet tall. Each plant can produce up to 20 stems from a vigorous crown and a deep tap root. The leaves are alternate, odd-pinnate, with six to ten pairs of leaflets. The tip of each leaflet has a small hair-like appendage. The stems are hollow and cylindrical, tubular. The stipules (leaf-like appendage at the base of the leaf stem) is sagittate (arrowhead shaped) and toothed and lobed. The pea like flowers are whitish to bluish to purplish, and are found in terminal or axillary racemes. Goatsrue produces 1-9 seeds per pod, and each plant can produce 15,000 pods per plant, or more. The seed pods are narrow, round in cross section, and about 1 inch long. The seeds are dull yellow, bean-shaped and 2 ½ times larger than alfalfa seeds. Seeds may be viable in the soil for 5-10 years, but testing continues on seed longevity.

Table. Comparison of Goatsrue and Wild Licorice

Characteristics	Goatsrue	Wild Licorice
Stems	hollow, cylindrical, tubular	solid
Seed Pods	narrow, smooth, >1 " long	bur-like with hooked
bristles, <3/4 "		
Leaves	with glandular dots when mature	

Economic Importance: Detrimental

Goatsrue is a federally listed noxious weed, with a very limited distribution, nationwide. One site is a county in Utah, that reports 38,000 acres (60 square miles) of infested cropland, irrigation waterways, pastures, fence lines, roadways and wet, marshy areas. Eradication efforts in Utah have been costly and time consuming.

Goatsrue is capable of forming a monoculture in wetland communities, displacing native or beneficial plants. Wetland wildlife vacate these areas once their food source or nesting material is gone. Goatsrue is fatal if ingested Goatsrue is unpalatable and toxic to sheep. Goatsrue is found in established alfalfa fields

Beneficial

Goatsrue is considered an ornamental species, and it is also considered a medicinal herb. Medicinal sources mention "veterinary and human medicine to increase lactation", and reduces blood sugar

Habitat

In Utah, goatsrue is found in cropland, ditch banks, irrigation waterways, uncut pastures, fence lines, roadways and wet, marshy areas. Goatsrue is also found in established alfalfa fields, where mowing does limit its spread. The King Co., Washington sites include roadsides, and an open field. The associated species growing with goatsrue in King Co. include reed canarygrass (Phalaris arundinacea), *Spirea*, blackberries (Rubus spp.) and purple loosestrife (Lythrum salicaria).

Geographic Distribution

Goatsrue is native to central and southern Europe and western Asia.

History

In 1891, goatsrue was introduced from the Middle East to Utah, where it was tested for three years as livestock forage, or a green manure crop. However, it is unpalatable and toxic (lethal) to sheep. Over the next 86 years, this plant slowly spread to cover a 60 square mile area in Cache County, Utah, where it is primarily found in highline canals and in drainage systems of valley floors.

In 1974 goatsrue was listed as a noxious weed in Utah, with the intent to keep this plant from spreading further in Utah, or into nearby states.

In 1976 goatsrue was the target of an eradication project in Utah (Evans 1984). In 1981 USDA/APHIS targeted goatsrue for eradication, nationwide. An herbarium search found plants from 1890 to 1960, from 10 continental states and Washington D.C. APHIS is working cooperatively with state agencies in Utah for total eradication.

Growth and Development

Goatsrue is a perennial legume. The flowers are present from June until the fall frost. The leaves and stems contain the alkaloid galegin, which is toxic to livestock when eaten in quantity. The alkaloid content is highest in the spring. Animals will avoid the plant, which contribute to the establishment and spread in rangeland.

Reproduction

Spreads by seeds. The seeds drop from the mature seed pods, and are spread primarily by irrigation or flowing waters. Seeds also spread by farm machinery, in animal manure and in contaminated soil movement. Seeds may be viable in the soil for 5-10 years, but testing continues on seed longevity.

Control

In Utah, an integrated approach, that included landowner education, carefully mapping the infestation, crop rotation, tillage, mowing, digging, hand clipping for seed pod removal and chemicals are used in the battle to eradicate goatsrue. "Massive reproductive crowns" are proving to be difficult to eradicate. Seed banks continue to produce seedlings, annually, in the original infestation area (Evans et al. 1997; Evans 1984).

Response to Herbicide

Selective herbicides are considered the most effective for large sites. Dicamba, or 2,4-D and their combinations are very effective. Clipping initial growth at 24" tall, and then spraying the regrowth at the same height is the most effective Be sure to check labels for site specific information on herbicide control before use.

Response to Cultural Methods

Alternative cropping and row crops are effective. Cultivation interrupts the life cycle of goatsrue.

Response to Mechanical Methods

Shallow cultivation, mowing, clipping and cutting are not recommended as a solitary control method. Flowers will be produced on very small plants. Seed pods can be clipped and disposed of to help prevent spread by seed in areas of eradication work.

GENERAL RULES OF SEED SOWING

Seeds need water and oxygen to germinate, so are best started in a light, loose soil that will not compact, get soggy, or crust over. Free flow of water & air are a must. Cover seed with 2 - 4 times their thickness of soil, unless they require light to germinate. Sow shallowly in cold wet spring, more deeply in warm dry summer. Large seed can be soaked overnight and planted singly. Barely cover small seed, and sprinkle fine seed on the surface and water by misting. Plant flat seed edgewise and winged seed with wing uppermost or broken off. Sowing too thickly wastes seed and weakens the crowded seedlings, but some kinds sprout best if crowded. Lightly tamp soil to insure good contact with the seed, unless heavy. Keep soil moist, not soggy, and do not allow to dry out.

Common causes of failure are soil too heavy, wet or cold, or allowed to dry out, not giving slow seeds long enough to come up, pests eating the seeds or seedlings, and not giving dormant seeds the proper pretreatment. Careful attention to the instructions in the catalog and on the packet will help insure good results. Common causes of seedling loss are damping off due to poor air circulation & overwatering, drying out or burning due to placing in full sun or outdoor conditions too quickly, transplanting shock (best done on a cool, moist day), and predation by insects, slugs and snails at night.

TEMPERATURE

Most seeds germinate best at warm (70°F) temperatures. Plants from temperate regions, the arctic, high mountains and high deserts often germinate best at cool temperatures. Plants from winter-rain areas like California, the Mediterranean, Chile, S. Africa and parts of Australia also like cool temperatures. Warm temperatures will often speed germination of these seeds, but lower vigor & survival. Warm desert plants and tropicals like warmth. Temperatures used in the catalog are: Cold (34 - 45°F), Cool (50 - 65°F), Warm

(65 - 80°F) and Very Warm (80 - 100°F). When no temperature is given, warm (70°F average) is indicated.

TIME UNTIL GERMINATION

Average range of time to germinate is usually given in weeks. A seed that takes 2 - 3 weeks will usually come up fairly evenly; one that takes 1 - 12 weeks will tend to straggle in irregularly. Time varies with temperature, so expect considerable variation. Don't give up too soon- many who have given up and sown another seed in the pot end up with two types of plants in the same pot!

HARDY ANNUALS (HA)

These will stand some frost and can be sown direct to the garden as early as the ground can be worked, March to June. Prepare the soil until a smooth, fine surface is obtained. An attractive annual border can be had by planting in large, irregular drifts.

HALF HARDY ANNUALS (HHA)

These are killed by frost and should be sown in late spring after danger of frost. For early bloom, start early indoors & plant out after danger of frost.

TENDER ANNUALS (TA)

These need warmth and shelter and are best started in pots or flats and planted out in favored spots after the soil has warmed.

BIENNIALS AND WINTER ANNUALS

Biennials are sown like half hardy annuals or perennials in spring or fall and planted out in September and October. Winter annuals such as some Californian and desert plants may be grown in summer, but are at their best sown in fall, even if grown in the greenhouse in cold winter areas.

HARDY PERENNIALS (HP)

Many of these germinate readily at warm temperatures, and can be sown direct to the garden or early in the greenhouse or cold frame. If started early, they often bloom the first year. Others germinate best at cool or cold temperatures and the seedlings need cool temperatures. Many have various dormancies & need the pretreatment indicated.

HARDY TREES AND SHRUBS

Many will germinate readily if sown in spring. Others need cold or other pretreatment, and some are best sown in fall or winter and covered with a mulch or snow. The addition of some forest soil or litter from below both hardwoods and conifers is often very beneficial, greatly increasing seedling growth due to beneficial mycorrhizal inoculation.

TROPICAL AND SUBTROPICAL PLANTS

These often need very warm temperatures. Bottom heat and constant moisture are often beneficial. Some take surprisingly long times to germinate. Many people think all tropical seeds are short-lived and perishable, but many of the longest-lived seeds are tropical, and some even need to be aged dry for a year before they will germinate!

AQUATIC PLANTS

We have had the best results with some of these by sowing in pots, pans, or petri dishes, and submerging so that the soil is under 1/4" of water.

FERN SPORES

Spores should be scattered thinly on the surface of a light, sterile, peaty mix which has been watered with clean sterile water about 4 hours before sowing. Cover pot with a pane of glass or enclose in a plastic bag and place out of direct sunlight. A light green film will form on the surface in 1 - 4 weeks or so, up to 6 months for slow species. *Botrychiums* may take up to 8 years, and one *Equisetum* takes 18 years! Small round or heart-shaped prothallia will then form. Begin to mist lightly with distilled water weekly to enable fertilization of the egg cells, until true fronds develop. Transplant sporelings when large enough to handle. Good ferns can usually be had in 8 - 12 months from spores. Ferns are hybridized by sowing 2 kinds of spores together.

SOIL MIXES

Many seeds do well sown direct to ordinary garden soil, but even good soil may be poor in pots or flats. These need a lighter, looser soil. Most commercial mixes are fine, but the addition of some garden soil and compost will often insure adequate beneficial micro-organisms & fungi, A good soil mix can be made at home from 1/3 garden loam, 1/3 peat or compost, and 1/3 gritty sand. Number 1 limestone chick-grit makes a good top dressing for many alpines or slow germinating seed to discourage algae growth. Crushed charcoal also helps.

HARD SHELLED SEED

These have hard impermeable shells and need nicking or scratching the coat to allow water to enter and the seed to germinate. The best results are from the least amount of nicking that will allow water to enter and the seed to swell. Many failures are due to over-nicking and damaging the seed.

Different seeds need varying amounts of nicking. Most do best with lightly rubbing on sandpaper or a file until just the very outer coat is scratched. Often just scratching with a knife-point or scriber works. Others need serious nicking, sometimes with a hacksaw until the white interior shows. Nicking seed one by one can be tedious but is most effective. Larger lots can be rubbed between two boards covered with sandpaper, or shaken or tumbled in a can lined with sandpaper. Then soak the seed overnight and any that don't swell or soften should be re-nicked and soaked again until swollen.

Nick and Soak means nick first, then soak, and re-nick any hard (unswollen) ones. Soak and Nick Unswollen means the seed is usually only partly hard and many will swell without nicking. Soak first, nick only the hard ones.

Other methods for large lots include:

Hot Water Soak: The seeds are placed in a cup and not quite boiling (200°F) water is poured on them and allowed to cool & the seeds to swell. At least 4 - 5 times their volume of water is used. Or the seeds are dipped in boiling water for 10 seconds to 3 minutes. Produces erratic results, often killing many of the seeds and increasing fungal attack.

Dry Heat: The seeds are baked dry in an oven at 140° to 220°F for 4 - 10 hours, or are microwaved for 30 seconds to 4 minutes. Gives erratic results. Sulfuric acid or lye soaks work with some seeds.

COLD TREATMENT (PRECHILL)

Many seeds need a cold moist period before they will sprout. The essentials are moisture, air, cold and time. Dry cold (a packet in the refrigerator) won't work. Often the seed may be sown in fall or winter and allow natural cold & snow to work.

- Soak seed overnight until swollen or soft (up to four days for large hard nuts). Nick if needed.
- Mix seed with about 3 times its volume of damp peat moss or vermiculite and place in a plastic bag or pill bottle. Small amounts may be conveniently layered between damp paper towels or coffee filters. Remember, air is essential; avoid sogginess. Label the bag with the name of the seed and date to be removed from cold. Mark your calendar too!
- Store in the refrigerator (34 - 40°F) for the time specified in the catalog. Four to 12 weeks is usual.
- Remove the seed and sow. Seed is best kept cool (50°F or so) for a week after sowing, and gradually brought up to warm temperatures. Warming too quickly can be fatal for some seeds.

'Cold germinators' only sprout at cold temperatures, and are most conveniently treated by layering in damp towels and pricked off as they germinate. If using outdoor treatment, hold pots at above freezing for a few weeks before putting outside. Snow cover is often highly beneficial.

Dormancy is highly variable. Sometimes a seed collected in the warm lowlands will germinate readily, but the same species collected at a high elevation will need cold. and varies with weather before harvest and conditions of storage after harvest.

GIBBERELLIC ACID-3 (GA-3)

GA-3 is a naturally occurring plant growth regulator. Presoaking seeds in GA-3 will often cause rapid germination of many highly dormant seeds that would otherwise require cold, aging, light or other prolonged treatments. Many different types of dormancy are overcome. You may buy our 'GA-3 Primed' seed which are ready to plant, or the GA-3 powder or one of our kits. GA-3 is safe, easy to use, economical and rapidly becoming a standard tool for germinating seeds.

OTHER TREATMENTS

Dry Storage or Ageing

Many seeds will not germinate when freshly harvested, but are dormant until after a period of dry storage ranging from 1 - 12 months or up to 5 years. The time varies with temperature, humidity and oxygen. We try to supply aged seed whenever necessary. Often this type of dormancy can be broken with GA-3.

Warm Moist Treatment

Many seeds need 1 - 4 months of warm moist treatment, followed by cold treatment to sprout. In some, the root sprouts during the warm period, but the shoot does not sprout until after a cold period. Called 'two-step' germinators by Deno.

Some of these have unformed embryos which must first develop at warm temperatures. Some seeds are '70 - 40' germinators, which need warm moist, then cold moist, and germinate at cold temperatures. Some tropicals which are slow to germinate (many palms and aralias) need prolonged warmth to first develop their embryos. Many seeds are 'multicycle' germinators, needing several cycles of cold and warm periods.

Soaking, Washing or Leaching

Many seeds contain germination inhibitors which in nature prevent the seed from germinating except during the wet season (tropicals) or only after sufficient rain has fallen (desert plants). Soaking the seed in shallow water and changing every day for several days will leach out inhibitors. Heavy daily watering of pots may work. Some have oily coats and need detergent or peroxide soaks.

Light or Darkness

Some seeds need light to germinate; others need darkness, and light prevents sprouting. If light is required, sow on the surface; if darkness is needed, cover seed well.

Oscillating Temperatures

Some seeds need the daily change from warm to cold that occurs in spring and fall, and they germinate best sown in pots and exposed to outdoor conditions.

Other Treatments

Soaking seed in potassium nitrate (KNO3), hydrogen peroxide, citric acid, sodium hypochlorite (bleach), smoke solution, charred wood leachate, or enzyme solutions have all been used to trigger germination of seeds.

SMOKE TREATMENT

Smoke treatment often helps germination of plants from fire-prone environments, particularly Mediterranean-climate plants such as many species from California, Chile, Australia, South Africa, and the Mediterranean region. For a list of genera that have responded to smoke treatment,

Prepare a dilute smoke solution by adding one part commercial smoke flavoring to nine parts water. Either soak the seeds in this solution overnight (or until they swell), or by water the pot or flat once with this solution. Smoke flavoring is found in the spice and flavoring or barbecue section of the grocery store . It comes in small brown bottles of liquid, called "liquid smoke" or "hickory seasoning". Look for the "all natural" type that lists only water and natural smoke concentrate as ingredients. "Wright's" is the brand we find out here in California.

Smoke treatment is still experimental, and you may have to try different dilutions.

PLASTIC BAG METHOD

Best for very slow to germinate seeds, very tiny dust-like seeds that can't be allowed to dry out, and very slow-growing seedlings. Small clean pots are filled with damp, sterile, soilless mix. Milled sphagnum or a light dusting with powdered charcoal discourage fungi & algae. Seeds are sown and the whole pot is sealed in a plastic bag and placed out of direct sunlight. This creates a mini greenhouse and the soil will not dry out and the seeds are protected from mice, etc. Pots can be left for years with no care other than regular checking for seedlings. As soon as seedlings appear, begin hardening off. Bagged pots may be kept under fluorescent lights without overheating. Don't forget to label!

HARDINESS ZONES

The rating of plants according to the hardiness zones in which they will grow has improved greatly over the last twenty years. They are fine for common, widely grown plants, but for most plants, our knowledge of their ecological limits is incomplete. Too often, the assignment of zones is merely educated guesswork. Many plants will do well far outside their usual ecological range, or seeds collected at the species northernmost limit will be much hardier than usual. Also, a heavy mulch, or planting by a south-facing wall, or other simple tricks can greatly extend a species range in cultivation.

SEED LONGEVITY

Many people have the mistaken idea that seed must be very fresh to grow, and the occasional failure of a packet is blamed on 'old seed'. Properly stored, seed of most species is viable many more years than most people would expect, some giving high germination for decades. Though some species are short-lived, many others are dormant when freshly harvested, and actually need an aging period of weeks, months, or even years before reaching full germination.

Seed *half-life* is frequently given in the catalog. This is the average number of years it takes for seed of that species to drop to half of maximum germination, when stored under ordinary conditions.

ABOUT DIFFICULT SEEDS

Over millions of years, wild plants have evolved germination strategies which ensure their survival, but which may not be convenient for the home gardener who wants a quick and even stand of plants from a packet of seed. Many seeds sprout irregularly, so that if the first flush of seedlings is killed by adverse weather, insect predation, etc., more will come along to take their place.

In adaptation to various environments, some seeds need periods of cold, warmth, darkness or light, fire, etc. Some have seedcoats of varying hardness or impermeability, and others contain chemical germination inhibitors which must be leached from the seed before it can sprout. Some species disperse themselves over wide areas by being eaten by animals, the seed sprouting far from the mother plant, the seedcoat softened by digestive juices. Many seeds have internal clocks, and give much higher germination at certain times of the year, regardless of the treatment given. All seeds wait for the correct time and conditions before sprouting, and the gardener must mimic those conditions to ensure successful germination.

We are continually testing seed for germination, and conduct research into improving methods of handling difficult species. For slow seeds, which take months or years, making a standard test impractical, we may use 2,3,5-triphenyltetrazolium chloride to test for hydrogenase enzyme activity, which quickly indicates whether there is a living embryo in the seed. It can take years of research to determine the best ways to germinate a specific species. *Oryzopsis* seed has been studied for over 50 years, yet we still do not fully understand its requirements. Even then, seed collected from one population may germinate readily, yet the same species gathered in a cold-winter area may need cold treatment.

About Hybrids

The word *hybrid* is used in two very different ways in the seed trade. First, to designate F-1 hybrids, and second, to designate open-pollinated plants which originated in the crossing of two distinct varieties or species.

F-1 means 'first filial generation', and *F-1 hybrids* are the first generation produced by crossing unlike parents, the offspring of which exhibit 'heterosis' (hybrid vigor) and are very uniform. The characters of F-1 hybrids are not stable, so that seed saved from F-1 plants will not come true, and may produce many distinct types in the second generation, often reverting to various ancestral forms. Therefore, the original cross must be repeated each year in order to produce seed.

The second meaning of the word is much older, going back to the last century. Many flower mixtures are called 'Hybrids' or 'New Hybrids'. These are *not F-1 hybrids*. They are the result of crossing unlike parents in order to produce variation, the progeny were then selected for desired attributes such as new flower colours, etc., and these characters fixed by selection. These are open pollinated, and are relatively fixed. The original crosses are not repeated - the various strains are kept pure by selection and isolation.

Open pollinated. or O.P., means that bees, wind, and the agencies of nature are allowed to pollinate the flowers, rather than by emasculating the flowers and applying pollen by hand as in F-1 hybrids.

I do not distribute F-1 hybrids. First, because the grower cannot save his own seed from them to multiply the plants as he chooses. This insures the gardener's and farmer's dependence on the seed company to produce and sell the seed to him each year. Second, gaze at a field of F-1 plants. They appear to be almost machine-made. This genetic uniformity of the first generation causes vulnerability to crop failure. This occurred in the wheat stem rust epidemic of 1954 which took 75% of the crop, and in the southern corn blight epidemic of 1970, which destroyed 20% of the crop. A famous example of disease wiping out genetically similar plants was the Irish potato blight of the 1830's, in which 2 million people starved, and 2 million fled, reducing the country's population by half. Third, the widespread cultivation of F-1 hybrids to the exclusion of the old open pollinated varieties narrows the genetic base of our crops, and contributes to the 'genetic wipe-out' which is alarming biologists and agronomists throughout the world. Finally, the methods used to produce the F-1 seed each year are inhumane, and contribute to the exploitive world-view which is destroying our environment.

An important point: in wild nature, hybridization between populations, races, varieties and species is a common, natural occurrence. In wild nature this is a highly beneficial process, promoting diversity and evolution. The objection here is to controlled, large-scale *industrial* production of hybrids, leading to uniformity rather than diversity, and leading to the corporate control of germplasm. There is no problem with natural hybridization, or with home gardeners experimenting with hybridizing plants in their backyards, except that each individual should give consideration to the ethical questions about the invasiveness and inhumaneness of the procedure.

2

Seed Quality Components

Forage seed quality depends on three key components: germination, genetic purity, and crop purity. The importance of each is described below.

GERMINATION

Germination is the measure of how well a seed produces a normal, healthy seedling when planted under ideal conditions. The standard germination test, evaluated under the temperature considered ideal for each species, provides an estimation of seedling emergence if soil conditions at planting and during early seedling growth are near optimum. The results of the standard germination test are printed on the seed tag

Each species has a specific range of temperatures at which germination occurs. If temperature at planting falls outside this range, germination may be slowed or prevented, and during extreme conditions the seedlings may die. Excessively high or low temperatures also can induce secondary seed dormancy, causing a delay in germination and establishment. It is therefore important for growers to be aware of the germination requirements of the species that is planted and to sow fields when conditions are favorable. Suggested planting dates usually reflect the requirements for specific species and are listed in *Production and Utilization of Pastures and Forages in North Carolina* t is essential that forage seed have adequate moisture immediately following planting to promote rapid germination and for a period of about three weeks after emergence to promote seedling growth.

GENETIC PURITY

Varieties are adapted for specific uses and production areas. The genetic constitution of a variety will influence disease and insect resistance, plant response to climatic extremes, forage quality, and many agronomic characteristics. Growers should purchase seed of a known variety that has been tested for their area. In North Carolina, forage seed may be sold *variety not stated*. However, growers should realize that purchasing seed of an unknown variety often leads to low yields, reduced stand life, and increased cost.Genetic purity, or trueness to variety, is established and maintained by special purification and seed increase programs, by field and seed inspections, and by pedigree records. The best assurance of obtaining genetically pure seed is to buy *certified seed*, clearly identified by the *"blue tag"* The blue certified seed tag is issued only to seed lots that meet specific genetic, physical, and physiological standards.

Crop Purity

Seed lots with even small amounts of weed seed or other crop seed can cause serious economic losses and diminish forage quality. Weeds and other crop plants compete with the desired species for nutrients, space, and soil moisture and are often difficult or expensive to control. Some weeds, such as serrated tussock, Canada thistle, or horsenettle, can harm animals or threaten pasture value. Other weeds that are frequently found in forage seed include dock, sedge, wild garlic, wild onion, mustards, panicums, and other forage seeds such as ryegrass.

SEED ANALYSIS

KIND AND VARIETY

The "kind" of crop refers to the species. North Carolina does not require the forage variety name to be listed on the tag. However, if the variety is not given, the seed tag should bear the statement, *variety not stated*. When two or more varieties or kinds are named, the word "mixture" or "mixed" must appear on the label. Bermudagrass, often called common bermuda, is no exception to these rules. The sale of "common" bermudagrass requires either the variety name or the notation *variety not stated*.

Germination

This notation tells the percentage of pure seed that germinates. The percentage is based on the number of seeds that produced normal seedlings following a specified test period (a normal seedling being one that has the essential structures necessary for plant survival). If more than one kind and variety are named, the germination percentage must be shown for each.

Some forage species are dormant when harvested, and the dormancy period may last for a few weeks to several months. When seeds are dormant, they are still viable but will not germinate when placed in the proper germination environment. Special pre-conditioning treatments must be used to break or overcome this dormancy. Switchgrass is a good example. Its seeds must have a pre-chill treatment for 2 weeks at 40°F to condition the seed for germination. Also, the seeds must be fully imbibed during the pre-chill treatment. When switchgrass is labeled 90 percent germination on the seed tag, it takes into account two weeks' pre-chill treatment during the test to break dormancy, followed by two weeks in alternating 60/85°F to measure germination potential.It is possible to buy seeds that have high germination potential, but if dormancy has not been broken, they will not germinate well when planted. Dormancy of most species can be broken by storing the seeds at room temperature for several months before planting.

Origin

This label entry identifies the state or country in which the seed were grown. If the origin is unknown, the statement *origin unknown* must appear on the label.

Lot Number

Lot number allows the seed producer and buyer to identify the specific unit from which the seed were taken. The term "lot" means a definite quantity of seed that is uniform throughout for the factors that appear on the label. This information is useful in cases of performance problems.

Net Weight

This is the weight of the seeds only – minus the weight of the container.

Pure Seed

This percentage tells the total weight of seeds of the kind and variety stated on the tag. If more than one kind or variety is named, the pure seed percentage of each component must be given.

Inert Matter

Extraneous material, such as dirt, stems, leaves, and seed parts in the seed lot, is described in this percentage. Inert matter reduces the value of seed. It is best to choose seed with less than 2 percent inert matter.

CONTRIBUTE TO SEED QUALITY

No single test is sufficient to distinguish all the attributes that contribute to the physiological quality of a given seed lot. From a growers view point, a high quality lot is defined by results. A high quality lot is one that gives fast, uniform, high percentage emergence, resulting in healthy, near 100 per cent stand under the conditions in the growers field or greenhouse. Unfortunately, defining quality by historical results is not enough. a predictive test of field performance is needed to provide growers a reasonable assurance of obtaining given good cultural practices and typical (or at least not greatly abnormal) weather, insect or disease conditions. The stand potential needs to be assessed both before and after coating. The goal for a commercial seed or coating company is to develop a fast, reliable series of tests that will predict accurately the stand establishment of a seed lot under a range of possible seed conditions, including stress.

The Federal Seed Act and various state laws stipulate that seed germination be labeled according to standardised tests established by the Association of Official Seed Analysts (Copeland 1981). For example, lettuce must be labeled with the percentage of germinated seeds counted after 7 days at a constant 20C in the light. Celery is tested under the same conditions but the percentage is counted after 21 days.

The official germination figure is extremely useful, but additional tests are needed to determine if the seed lot is suitable for coating. This single laboratory test does not provide sufficient information to determine if the seed lot will retain its germination pattern (percentage, speed, uniformity, reaction to high or low temperatures and shelf life) after coating or ho it may perform under actual conditions in growers fields or greenhouses. Additional tests can include variations on the standard lab test, thermogradient table tests and greenhouse tests.

LABORATORY TESTS

Variations on the standard germination test increase the stress level, usually using temperature and/or darkness. Lettuce might be tested at 15C or 25C or alternating at 20/30C, perhaps in the absence of light. The lettuce variety Etna was rejected for coating based on such a test, while Domingos 43 was accepted (Table). Etna failed to meet the coating companies standard of > 93 per cent germination with a < 3 per cent abnormal seedlings at 20C and was not coated.

Time	Percentage			
	Raw Seed at 20C	**Raw Seed @ 20/30C Coated Seed**	**Coated Seed at 20C**	**Abnormal Seeds**
Etna lot 1292				
1 day	89	90	–	* % abnormal seeds
7 days	92 + 7*	92 + 2*	–	* % abnormal seeds
Domingos 43 (lot 5980)				
1 day	100	68	99	* % abnormal seeds
7 days	99 + 1*	99	98 + 1*	* % abnormal seeds

This standard is empirical, not theoretical, and was chosen based on years of experience with lots that give satisfactory or unsatisfactory field results. The standard is conservative and if no other lot is available,, the seed might still be coated provided that the grower knows, and explicitly accepts the limitations of the seed and the risk of a weak stand. In another test, the onion variety Sweet Perfection tested as an excellent raw seed lot (Table). Germination patterns for onions differ from lettuce; onion seed lots with total potential 80 per cent to 85 per cent frequently are encountered and accepted by growers. Lettuce seed lot germination < 90 per cent are strongly resisted by commercial growers, who prefer seed with a germination of > 95 per cent.

Table. Germination per cent of Raw and Coated "Sweet Perfection" Onion Seed. Lot 57412, after Two Time Periods in the Lab at 20C and Four Time Periods in the Greenhouse with no Temperature Control.

Time	Percentage		
	Raw Seed at 20C	Coated Seed	Abnormal Seeds
	Laboratory at 20C		
3 days	74	83	* % abnormal seeds
7 days	94 + 2*	95 + 2*	* % abnormal seeds
	Greenhouse, no Temperature Control		
6 days	–	18	* % abnormal seeds
8 days	–	82	* % abnormal seeds
10 days	–	95	* % abnormal seeds
14 days	–	95 + 2*	* % abnormal seeds

Thermogradient Table Tests

A Thermogradient table is a uniform sheet of metal with a cold water bath along one side and a hot water bath along the other. This creates a continuous temperature gradient, typically with a range of 20C. *e.g.*, 12C to 32C or 16C to 36C. ten individual tests can be run on blotters across the width of the table, at 2C increments. This type of test determines the maximum and minimum temperatures at which an individual lot will germinate, as well as its ideal, or preferred temperature range for fastest germination and growth The different tests are used in combination. For example the lettuce variety Domingos 43 showed heat dormancy at 20/30C, with reduced germination after 24h (Table). The severity of the heat dormancy was tested on the thermogradient table. The test indicated a potential germination of >87 per cent in 24h and 96 per cent in 48 h in temperatures up to 23C. This is acceptable for this variety at the usual soil temperatures when it is sown in Southern California and Arizona.

Greenhouse Tests

Greenhouse tests give a stand establishment estimate under real conditions. These tests also can be prolonged to give a subjective estimate of seedling quality. The greenhouse test can, for example, identify seedlings with necrotic tissue on the cotyledons that might not be apparent in other tests. The percentage of seedlings with significant necrosis is deducted from the stand establishment potential of the lot.

Seed quality testing is an integral part of seed coating. Without it the performance of coated seed would be erratic and unreliable for commercial growers. Precision sowing without dependable, precise emergence wastes the advantages of coated seed. For example, the Etna probably would have a field or greenhouse stand well below 90 per cent. Depending on circumstances, this might be acceptable, but it would fail to meet the expectations of many growers.

No single test is adequate to accept a raw seed lot for coating, but several in combination can identify acceptable and unacceptable lots with a high level of accuracy. Prudent coating companies repeat all or part of their battery of tests on the final coated product to ensure that the commercial product meets the expectations of the pre-coating tests. Testing methods, which differ among companies, have been developed to evaluate coatability of a raw seed lot and to check the coated lot's quality. This intensified testing for subtle quality distinctions has improved understanding of quality and how to measure it. This helps both seed and coating companies to improve quality with techniques employed in seed growing, harvesting, conditioning storing and coating. When it is possible to measure the quality differences it is also possible to develop and choose superior techniques. The result has been a gradual increase in quality standards throughout the seed industry. Thus, "The Stimulus to Seed Quality" in the title of this article. For example 15 years ago, lettuce could be sold at 85 per cent germination and 90 per cent was considered good. Today, 90 per cent is considered a weak stand and growers seek 95 per cent or better. Commercial growers have higher expectations of physiological performance, mirroring their higher expectations of genetic performance by new varieties of each crop species.

This developing ability to measure and deliver higher seed quality reinforced by the markets higher expectations, is the major long term benefit of seed coating. The particular economic benefits of precise placement, reduced thinning costs, ease of handling etc., motivate growers to change from raw to coated seed and pay the associated research and production costs. Meanwhile, the on going pressure to maintain and exceed existing quality standards benefits all participants.

ENHANCEMENT OF SEED QUALITY

Oilseeds, the raw material for vegetable oil, occupy a significant place in India's economy. Next to food grains, oilseeds account for 10 per cent of the cultivated area and value of all agricultural produce. Nearly 85 per cent of the oil and fat needs of the country is primarily met by vegetable oils. India is the third largest producer of oilseeds in the world. No other country has its range of perennial and annual oilseeds. In terms of area, India ranks first in groundnut, sesame, linseed, safflower, niger and castor. Although India has 20.8 per cent of the world's area under oilseeds, it accounts for less than 10 per cent of world's production.

In terms of vegetable oils, India is the fourth largest oil economy in the world after the U.S, China and Brazil. Favourable agro-ecological conditions in the country have supported commercial cultivation of seven annual edible and two non edible oilseed crops, besides a number of minor oilseeds of horticultural and forest origin, including in particular coconut and oil palm. With the growing population along with raising per capita income, the demand for vegetable oil is likely to grow unabated in the coming years. The impact of green revolution has made our country self sufficient in the production of cereals. The situation so far as oilseeds are concerned is not very bright. Although India is one of the world's largest producer of oil seeds, the quantity of edible fat available falls far short of the country's requirement.

Seed being the basic input in agriculture, production and supply of quality seeds to the farmers will go a long way to achieve the goal of self sufficiency in oilseed crops. Very often the seed may retain the required germination standard but its vigour, which is yet to be quantified and productivity. Seed pelleting or coating is the most applicable technique in direct sown crops which need initial vigour for sustained crop growth and development.

Coating provides an opportunity to package effective quantities of materials such that they can influence micro environment of each seed which supplies not only micro and macro nutrients but also protects the crop from pests and diseases from the earlier stages due to the inclusion of Pesticides. Because of increased awareness of organic farming, use of plant products in agricultural research is gaining importance. Seed coating is an important area in which plant materials could be widely used. To increase the productivity, various presowing seed treatment methods are followed but such treatments have not come into practice among farmers because of the non-availability of the chemicals to the marginal and small farmers and they are also expensive. Hence, an attempt was made to explore the use of easily available materials for pelleting to increase germination and seedling growth in early phase. Researchers are recommending several pre-sowing seed management techniques with the benefit of invigouration, protection and production Among the various seed treatments, the physical seed treatments *viz.*, pesticide seed treatment, seed colouring, seed coating and seed pelleting are widely under the usage by the seed producers and are normally included as a continuous treatment in sequencing the various post harvest seed handling techniques.

Basu revealed that the seed treatment implies an improvement in seed performance resulting in better field performance than the corresponding untreated seed and the establishment of a seedling in the soil is an important and foremost need for the better crop production. This depends largely on the germination and vigour potential of the seeds used for sowing. Seed treatment provides an opportunity to package effective quantities of materials such that they can influence micro environment of each seed which supplies not only micro and macro nutrients, but also protects the crop from pests and diseases from the earlier stages due to the inclusion of pesticides. To increase the productivity of the oilseeds, various physical seed enhancement techniques were followed but such treatments have not come into practice among farmers because of the non-availability of the chemicals to the marginal and small farmers and they are also expensive. With this background, experiments have been conducted to study the influence of physical seed enhancement techniques such as halogenation, fungicidal and pelleting on productivity of sesame cv VRI 1, sunflower cv Morden and groundnut cv VRI 2 kernels.

MATERIALS AND METHODS

Genetically pure seeds of sesame (Sesamum indicum) cv. VRI 1, Groundnut cv. VRI 2 and Sunflower cv. Morden obtained from the Oilseed Research Station, Virudhachalam and Agricultural Research Station, Bhavanisagar formed the base material for the study. The bulk seeds were graded for uniformity using appropriate round perforated metal sieves of sizes *viz.*, 5/64" for Sesame, 9/64" for Sunflower and 18/64" for Groundnut kernels. The field experiments were conducted at the Department of Agricultural Botany, Annamalai University, Annamalai Nagar (11°24'N latitude and 79°44'E longitude with an altitude of +5.79 mts above mean sea level).

The seeds of above oilseed crops were imposed with the following physical seed enhancement techniques.

- T0 - Control.
- T1 - Dry dressing with Thiram @ 2 g/kg of seed.
- T2 - Slurry treatment with Thiram @ 2 g/kg of seed.
- T3 - Dry dressing with Halogen mixture @ 3 g/ kg of seed for sesame and sunflower and 4 g/kg for groundnut.
- T4 - Slurry treatment with Halogen mixture @ 3 g/kg of seed for sesame and sunflower and 4 g/kg for groundnut as slurry.
- T5 - Pelleting with Arappu leaf powder @ 200 g/kg of seed.
- T6 - Pelleting with Neem leaf powder @ 200 g/kg of seed.
- T7 - Pelleting with Pungam leaf powder @ 200 g/kg of seed.
- T8 - Pelleting with Vasambu rhizome powder @ 200 g/kg of seed.

The treatments were evaluated for field stand, plant height, number of branches, pod yield, seed yield, seed weight, seedling length, dry matter production, for all the three crops. In addition to the above characters, capsule morphological characters for sesame, head morphological characters for sunflower and pod morphological characters for groundnut were also observed. The experiments were conducted for two consecutive years, their average values were worked out and subjected to statistical analysis. The physiological parameters such as chlorophyll content, leaf area index, relative growth rate, net assimilation rate, germination per cent and vigour index were also evaluated using following methods. Chlorophyll content.

A quantity of 250 mg of leaf tissue (third leaf from top) was homogenised with 80 per cent acetone and centrifuged at 3000 rpm for 10 minutes. The supernatant was collected and made upto 250 ml with 80 per cent acetone.

The optical density of the extract was read in a spectronic 20 using 652 nm for total chlorophyll using 80 per cent acetone as blank. The total chlorophyll content was computed using the formula and expressed as Mg/g of fresh leaf.

SEED ENHANCEMENT

There is obvious value to improving performance beyond that of raw seed Techniques to evaluate seed quality have been refined further to evaluate methods of enhancing performance.

Enhancement techniques can include seed production and conditioning methods, addition of chemical and biological agents, and seed priming. The purpose of these techniques include increase total germination, faster germination, better uniformity and temperature range of germination, to break post harvest dormancy, to reduce light sensitivity and to prevent disease.

Many of these techniques can be combined with coatings. The "Royal Green" lettuce, like nearly all green leaf lettuce, can be a slow irregular germinator and/or the absence of light (Table).

Time	Percentage			
	Raw Seed at 20C, dark	**Raw Seed @20/30C, light**	**Primed and Coated @ 20/30C, light**	**Abnormal Seeds**
1 day	98	98	99	* % abnormal seeds
7 days	98	93 + 1	98 + 2*	* % abnormal seeds
Primed and coated seed, no temperature control, light				
4 days		85		* % abnormal seeds
6 days		99		* % abnormal seeds

Note: * Raw seed is not greenhouse tested for this product because only the primed, coated seed will be sown commercially. The tests on raw seed evaluate coating suitability; the primed, coated product receives this final; test to ensure that the final quality has been achieved.

Low light conditions make such varieties especially prone to heat dormancy at relatively cool temperatures. Priming the seed can reduce the need for light and simultaneously improve the range of temperatures where germination is rapid and uniform. The thermogradient table test, in the dark, is indispensable for quality evaluation of primed and coated product illustrated.

Fungicides are another example of an enhancement applied to seed to protect vulnerable seedlings from various fungal diseases. Dust or slurry dithiocarbamate treatments are used widely and generally are successful. However, dosage is variable from seed to seed, and there is a degree of "dust off" when the seed is handled.

Fungicide treatments can be combined with film coatings and thereby applied at even more dosage rates, simultaneously eliminating dust off for a cleaner safer product. Powder coatings achieve similar results by blending precise amounts of fungicides in the coating powder for nearly identical dosage on each seed. Then a final layer of coating powder without fungicide can be applied at the end of the coating process, eliminating the chemical from the pill surface.

The preceding paragraphs are meant only to hint at the exciting potential of existing rapidly developing new products for enhancing seed performance. Secure production of excellent stands, even under stress conditions, is a worth while goal for researchers, the commercial seed industry and growers.

COATINGS

Coatings first were developed for cereal seeds in the 1930's by Germains, a British seed company. Large scale commercial use of coating began in the 1960's with precision sowing for the European greenhouse transplant industry. When California outlawed the short handled hoe in the mid 1970's, the use of coated seed for precision field seeders increased.

Precision sowing greatly reduced the number of skips and doubles in soil blocks or cell trays for transplants. Field precision sowing spaced seeds and thus, individual seedlings sufficiently to permit accurate thinning with a long handled hoe, while reducing damage to the root systems of remaining plants. The combination of U.S., field precision seeding and greenhouse transplant production created a demand for high quality coatings to achieve accurate sowing, satisfactory seedling emergence and stand establishment. University and commercial research programmes responded to this demand, producing coatings now widely used for seeds of vegetables, flowers and some field crops. (Kurosawa, 1976; Mayberry and Robinson, 1982; Markey,1990; Robinson et a, 1983; Valdes and Bradford 1987; Bradford *et al*, 1985)

In the United States, the major high-volume vegetable crop being using seed coatings is lettuce. Roughly 95,000 Ha (235,000 acres) are sown with coated seed. Brassicas, carrot, celery, endive, escarole, onion, pepper and tomato also are coated to a significant extent, varying with growing season, individual grower preference, the use of direct sowing or transplants, the economics of seed and coating costs etc. Begonia is the flower crop most frequently sown in coated form. Impatiens, marigolds and petunias are also coated commercially and the market for coating these and other species is growing strongly. Alfalfa and tobacco are two agronomic crops that are coated.

CONCEPT

Seed is coated when growers need a precision-sown crop and the non-coated ("raw") seed is too small, light or variable in size or shape to be sown accurately with existing equipment. Precision sowing is desirable when growers need singulation, *e.g.*, for cell-tray plant production in a greenhouse or strict control of spacing or depth of placement (*e.g.*, onion spacing is critical to achieve desired bulb size at harvest) Singulation and controlled spacing are also vital for crops that are direct sown and then thinned back to the desired population. The field thinning operation is faster, cheaper and more accurate when coated seeds are used. One Florida grower who used raw seed before 1984 informed me in 1985 that reduced thinning costs paid nearly all of the additional cost of coated seed. He stated that the stands were superior, and that this improvement was essentially "free" Superior stands meant that the incidence of skips or doubles was reduced and the plants were spaced closer to the ideal of 11.5 to 12 inches on center in the row.

The objective of coating is to deliver the seed in a form that is larger, rounder, smoother, heavier and more uniform than the original seed. The coated seed can then be sown with a belt, plate, cup, vacuum or other type of seed. The coated seed or "pills" can be placed individually, with improved spacing and depth control. The pills also flow better through the seeding mechanism, because their surface is smoother than that of non-coated seed.

Coating can be a carrier. Fungicides and beneficial microbials that protect the seed and emerging seedling are carried in the coating. For example, alfalfa seed coating with incorporated rhizobacteria is used to inoculate the field with beneficial microbial.

PROCESS

Seed coating relies on technology developed by the pharmaceutical industry to make medicinal pills. Commercial seed coating operations put seed in a rotating pa, mist with water or other liquid and gradually add a fine inert powder, *e.g.*, Diatomaceous earth, to the coating pan. Each misted seed becomes the center of an agglomeration of powder that gradually increases in size. The pills are rounded and smoothed by the tumbling action in the pan, similar to pebbles on the beach. The coating powder is compacted compression from the weight of material in the pan.

Binders often are incorporated near the end of the coating process to harden the outer layer of the pill. Binders can also reduce the amount of dust produced by the finished product in handling, shipping and sowing. Care must be taken with binders to avoid delaying or reducing the germination percentage.

Specific details of the materials used as binders are closely held as proprietary information by the coating companies. We are unaware of any public information on the classes of materials used as binders. Blanks and doubles are eliminated by intensive screening and other techniques. Uniform size and uniform rate of increase in size are evaluated throughout the process with frequent hand screening. At intervals during coating, and at the end, all of the pills are removed and mechanically sized on a set of vibrating screens. Smaller pills are returned to the pan and built up to the size of the remainder of the lot. After drying, usually with a forced air system at controlled, moderate temperatures, the pills are screened a final time before packaging. Undersized pills may be built up or discarded. The recovery rate (number of pills divided by the original number of seeds) has been 97 per cent +/–2 per cent for commercial seed lots at one commercial company for the past 10 years.

Species	Seeds / lb (Raw)	Pills / lb (Coated)	Weight Increase (%)
Begonia	70,000,000	250,000	28,000
Lettuce	430,000	12,500 - 40,000	3,400 - 1,100
Onion	120,000	15,000 - 30,000	800 - 400
Tobacco	5,000,000	165,000	3,000

Size uniformity after coating is a major criterion of coating quality. The usual tolerance for coated seed is +/–1/64th inch (0.4mm). This is the US seed trade standard for sizing, established long before coatings were introduced. For example, coated lettuce seed is sown most frequently with a belt planter through a 13/64 inch diameter round holes in the belt. This hole size requires that the lettuce pills be sized over a 7.5/64 inch screen and through an 8.5/64 inch screen. These tolerances result in levels of singulation well above 95 per cent in the field with placement in the row controlled within < 1/2 inch.

Accuracy of seed placement can vary with weight of the pill as well as the size tolerances. Sowing accuracy also depends on the skill of the equipment operators, the adjustment and the wear of the seeder, and the speed of the tractor through the field. The same constraints are true for greenhouse seeding: experience, attention to details and appropriate equipment are necessary to obtain the full benefit of coated seed.

TYPES

Two basic types of pill produced with inert coating powders are dissolving or "melt" coats and "split" coats. The melt coats dissolve when wet and gradually wash away from around the seed. Split coats initially retain their shape when wet and, by capillary action, pass moisture through the pill to be imbibed by the seed. The seed swells and cracks the pill by internal turgor pressure. The split coat often permit germination with less water, as they split, allow uniform, rapid oxygen access to the surface of the seed. The melt coats often require more water to wash the coating material away from the seed, and more time for the oxygen to reach the seed through the saturated coating material. Melt coats may offer advantages when soils are saturated, but oxygen availability always influences the speed, uniformity and total percentage of germination.

Powder coatings, both split and melt, multiply raw seed rate and depending on the coating, the number of seeds per pound may decrease dramatically. Generally, the heavier pills are easier to because they bounce an roll less than the lighter pills. However, the sowing equipment must be able to handle the pill weight. This can be critical, especially for vacuum type seeders.

In addition too the types of coating products described above, there is recent and increasing use of "film-coating". A thin polymer film smoothes the surface of the seed for better flow ability. The polymer also influences water uptake and the adherence of chemical fungicide treatments. Film coating only increases the raw weight of the seed 1 per cent to 5 per cent, far less than the powder coatings. Seed coating aims to influence the external physical properties of the seed, affecting the sowing characteristics only. By itself an ideal coating would be neutral in its influence on the speed, uniformity and percentage of germination when compared to the original raw seed lot. The ideal seed coating would perform in the same manner as the raw seed under a wide range of environmental conditions: light, moisture, temperature, pH, soil type etc. Also the stress of the coating process should not influence the germination pattern or longevity (shelf life) of the seed lot adversely, nor induce secondary dormancy — *i.e.*, affect seed quality.

OTHER CROP SEED

The percentage of the total weight from seed from a crop other than the kind and variety listed is indicated here. High-quality seed should contain no seed from other crops. The presence of other crop seeds in forage grasses can be particularly troublesome. Annual ryegrass in fescue is considered "other crop seed," not weed seed. The presence of 0.05 percent ryegrass in a forage translates into as many as 120 ryegrass seeds per pound of fescue. At a planting rate of 15 pounds per acre, this could mean as many as 1,800 ryegrass plants per acre of pasture.

WEED SEED

Weed seed contamination is expressed as a percentage of the total weight. This classification includes seeds, bulblets, or tubers of plants recognized as common weeds by official regulations or by general agreement. Weed seeds are typically very small, so a small percentage of contaminants can translate into many weed plants. High-quality seed should contain less than 0.2 percent weed seed.

Seedsman or Vendor

The name and address of the person or company labeling the seed are given on the seed label. They are responsible for the accuracy of the label.

Hard Seed

Hard seed are viable, but have a seed coat that is impermeable to water. Where hard seeds are present, total germination percentages are customarily determined by combining germination and hard seed percentages.

Test Date

The law requires that the germination test must have been made within nine months of the date the seed are sold or offered for sale (not counting the month of the test). The law also requires the individual owning the seed to keep the germination test date current, so carryover seed should be retested and re-labeled before the original test date expires.

Treated Seed

Seeds that are treated with chemicals must be labeled to show the chemical used, and the words "Caution," "Poison," or "Poison Treated" must be written on the label, depending on the harmfulness of the seed treatment to humans. In addition, the statement, "Do not use for food, feed, or oil purposes," also must be printed on labels of treated seed, and the seed must be dyed a bright colour.

Endophyte

The N.C. Seed Law does not require labeling for endophyte in tall fescue. However, if the seed are labeled, the N.C. Department of Agriculture and Consumer Services (NCDA&CS) does check to ensure levels of endophyte are correctly stated. If a discrepancy is found, a "stop sale" will be placed on the seed lot.

Noxious Seed

The name and number of all noxious weed seed must be listed on the tag.

DRESSING AND PELLETING OF SEED

In seed dressing or seed pelleting treatments, the surface of the seeds is coated with some inert material, the sticker, to which chemicals of various kinds may be added.

Some of the advantages claimed for pelleting and listed by Magini are:

- The incorporation of fertilizer material in the pellet furnishes the young germinated seedlings with needed nourishment.
- Plant growth regulators or stimulants may promote rooting or hasten emergence of the seedlings.
- Fungicides and insecticides are more effective when in direct contact with the seeds.
- Seeds may be protected against rodents by adding unpalatable, repellent or poisonous substances.
- Small seeds become larger and heavier and this improves the scatter pattern in aerial seeding.

Pelleting can also be used to incorporate a brightly coloured substance with the seed, which renders it more visible against the soil and thus facilitates uniform sowing rates. Coating with an anti-desiccant material such as alginate can preserve the seed against drying out, until adequate rain has fallen to permit safe germination.

Investigations have shown that some fungicides used in seed dressings may have an adverse influence on germination. The adverse effect was found to increase with temperature and this calls for extra caution in tropical countries. The cost of seed dressings is seldom justified when plants are raised in a nursery. Fertilizers, fungicides or insecticides can be applied more conveniently to the nursery soil than to the seeds, while good nursery hygiene coupled with presence of nursery staff through the day and periodic trapping or poison-baiting should be effective against birds and rodents. Pelleting is, however, sometimes used to improve the uniformity of seed for precision sowing in the nursery.

The main use of pelleting is for direct seeding, including aerial seeding. Protective measures to assist individual seeds after sowing are impracticable and pelleting is the only possible means of achieving some degree of protection. Most emphasis is now given to seed protection through incorporation of fungicides, insecticides and repellents; inclusion of fertilizers is seldom practised. One example of successful use was in aerial seeding of pines in the southern USA. The coating formulation consisted of endrin and arasan as the protectants, with a latex sticker to act as a bond. The seedling yields in field studies comparing coated with untreated seed were 55 to 1 for Pinus palustris and 12 to 1 for P. taeda. Seed coating of eucalypts is practised for aerial seeding of cutover mountain forests in southern Australia, where an area of 8,000 to 12,000 ha annually has been regenerated by this method. Aerial seeding has been used very little in the tropics but some success was obtained in Indonesian trials in Central and Eastern Java on areas dominated by Imperata grass, using Leucaena leucocephala,Calliandra calothyrsus and Acacia auriculiformis.

MATERIALS AND METHODS

Latex, methylcellulose and hydrol emulsions are used as stickers. Flaked aluminium powder is sometimes incorporated, because its shining quality is effective in scaring off birds, or to hasten drying and prevent clumping of the treated seed. A variety of chemical protectants is available.

Thiram or arasan, endrin and anthroquinone are among those commonly used. Thiram has been found effective against damping off. Red lead was long used as a dressing to protect seeds against birds and rodents but it is now considered to be valueless in this role. It has reduced germination and growth in Larix and may sometimes be toxic to the operator in hand-sowing techniques. A more suitable dye which is equally effective as colouring and which has no apparent adverse effect on subsequent germination is a Waxoline dye such as Lithofar red or the later Waxoline red A. It can also be used as a marker for special certified seed batches in storage.

Magini describes one method of coating seeds through use of a small cement mixer. The seed is placed in the mixer (a batch of about 12 kg at a time is suitable) and moistened with a latex sticker solution of one part latex to nine parts of water at the rate of one eighth to one quarter of a litre per kg of seed. Enough treating powder is then added to dry the sticker, usually in a volume relationship of four parts powder to one part sticker. The thickness of the seed coating is dependent on the amount of sticker in relation to the amount of seed. The total mixing time should not exceed four minutes, since prolonged agitation damages seeds or chips off the pelleted coat. The procedure described by Aldhous for coating seed with a Waxoline red dye is first to prepare a mixture of 1 part of dye to 19 parts of French chalk.

The germinating seeds of IDS/PREVAC-treated Pinus caribaea and P. oocarpa after 7 days. Note development of fungi on series with dead seeds. Fungi were much more abundant at end of test (21 days).

In the Philippines tests have been made of Arasan as a rodent-repellent seed coating for aerial seeding of Leucaena leucocephala. A sticker solution consisting of 1 part Dow Latex 512 R and 9 parts water was added to the seeds and the mixture stirred for 2 minutes. Arasan 75 wettable powder was then added gradually and the mixture stirred for a further 4 minutes, then spread out to dry for 14 hours. Several different strengths of Arasan were compared with a control. It was found that Arasan did not effect germination adversely, in fact the 7.5 per cent active ingredient Arasan gave significantly better germination than the control (74 per cent compared with 64 per cent). Results from 5 per cent and 10 per cent acid equivalent Arasan were intermediate. Previous trials had demonstrated that Arasan was effective in reducing rodent damage.

US Forest Service recommendations for preparing repellent and applying it to Pinus elliottii seeds are illustrated in Figure. In the southern USA 41,000 ha were direct seeded with pines in 1977 and 32,000 ha in 1978.

In Honduras the following mixture has proved suitable for Pinus oocarpa and P. caribaea direct sown: 60 g Arasan, 20 g 50 per cent Endrin, 5 ml latex and 100 ml water for 1 kg of pure seed. This concentration is considerably lower than that recommended for southern pines in the USA, since it was found that the higher concentrations caused some damage to these tropical species (Robbins 1983a, 1983b).

OTHER TYPES OF PRETREATMENT

One pretreatment designed to separate viable full seed from non-viable full seed has been described by Simak. It involves a form of pregermination of the sound seed and is thus a treatment suitable for application between storage and sowing and not between processing and storage. It is appropriate in cases where seed must be sown after a long period of storage, either because of a long interval between good seed years - a decade or more sometimes in northern conifers - or because of storage for conserving gene resources. Seed stored for such long periods may contain a substantial proportion of full seeds which have lost the power to germinate and the separation of non-viable from viable seeds greatly facilitates subsequent nursery operations.

The method involves three stages:

1. The provision of conditions ideal for initiating the internal seed processes which lead to germination in viable seeds, especially the imbibition of water
2. Partial redrying of the seeds. The viable seeds retain more absorbed water after redrying than the dead filled seeds and their specific gravity is therefore greater.
3. Separation, *e.g.*, by flotation, of the non-viable from the viable seeds. In one seed lot of Pinus sylvestris the treatments were
   - 16 hours' soaking in water, followed by 72 hours' storage in an incubator in a 2 cm thick layer at 15°C and 100 per cent moisture
   - 12 hours' drying at 15°C, 35 per cent RH and light of 800 lux
   - Separation into two fractions of "floaters" and "sinkers" by flotation in water.

This method, applied to 10 kg of seed of 67 per cent germination, separated the seeds into one fraction of 7.3 kg "sinkers" with 90 per cent germination and a second fraction of 2.7 kg "floaters" with 13 per cent germination. The resulting component with the higher germination can be sown at the rate of a single seed per paper pot in the nursery and germination is more rapid because of the pregermination treatment.

The method, known as the IDS (Incubation-Drying-Separation) treatment, requires very close control of temperature and moisture conditions if the desired differentiation between live and dead seeds is to be achieved without bringing the pregermination effect on the live seeds to the stage of radicle emergence. The optimum times of treatment may vary from seed lot to seed lot and it is necessary to carry out a preliminary trial on a small sample of each lot before treating it in bulk.

The method is likely to have most application in circumstances where:

- A substantial proportion of full but non-viable seeds is expected after storage
- Close control of temperature and moisture conditions is possible
- The viable seed obtained after separation can be sown in the nursery with minimum delay.

It is also possible to separate full but mechanically damaged seeds from full undamaged seeds. One method described by Lestander and Bergsten (1982) consists of placing the dry seeds in a drum partially filled with water. Centrifugal pressure is applied to the system by revolving the drum (5000 revolutions per minute has proved an effective speed) for a period sufficient to cause the damaged seeds to absorb enough water to make them sink, while the undamaged seeds absorb very little water and remain floating. In a seed lot of Pinus sylvestris known to have 26 per cent mechanically damaged seeds, it was found that 20 per cent of the seeds had sunk after 1 minute of drum rotation and 25 per cent after 5 minutes. About 98 per cent of the sinkers were damaged and only 2 per cent of the floaters. Germination in the sinkers was about 3 per cent and in the floaters 85 per cent (after 5 minutes' treatment and 21 days' germination test period). The same effect can be obtained by applying pressure to the liquid directly instead of by centrifuging, or by first applying a vacuum and then releasing it. The method is known as the PREVAC (Pressure-Vacuum) method.

This treatment is based on the fact that seeds with undamaged testas absorb water less readily than those with damaged testas, whereas the IDS treatment is based on the fact that a live imbibed seed releases moisture less readily on drying than a moistened dead seed.

The PREVAC and IDS methods have been successfully used in combination to improve seed quality in Pinus caribaea and P. oocarpa (Simak 1984). In P. caribaea the combined treatments increased germination after 21 days from 75 per cent in the control to 87 per cent (and to 91 per cent after a second incubation period), in P. oocarpa from 93 per cent in the control to 99 per cent. Speed of germination after sowing was also improved from 4 per cent after 7 days to 35 per cent in P. caribaea, and from 6 per cent to 58 per cent in P. oocarpa. The schedules used were those devised for P. sylvestris and further improvement in results may be obtainable by adapting these to the tropical pines. In the initial experiment about half of the seeds separated as "mechanically damaged" or as "filled dead" actually germinated in P. oocarpa and about one third in P. caribaea; on the other hand a much higher proportion of these "rejected but germinated" seeds showed signs of abnormal germination or reduced vigour than in the control.

3

Seed Cleaning Methods

The main characteristics by which sound seeds may be distinguished from inert matter including sterile and empty seeds are size and shape, specific gravity, colour and surface texture.

The ease with which sound seeds can be differentiated depends on:

- The degree of difference which exists between the seeds and the matter to be separated from them
- The degree of uniformity among the seeds themselves.

Colour, size and shape are useful criteria for visual separation, while most seed cleaning machines make use of seed size and specific gravity. Screening and sieving methods separate by seed or particle thickness or diameter; the indented centrifugal cylinder by particle length; liquid flotation and blowing, fanning and winnowing methods by specific gravity; while frictional cleaning methods rely on differences in surface texture.

Modern cleaning machines often combine more than one method, so that the cleaning process is both effective and quick. However, the species and the amount of seed to be handled will determine whether cleaning is best carried out by hand, by improvised equipment or by specialist machinery. The following account of cleaning and grading methods is based on Turnbull.

SCREENING OR SIEVING

In most cases a number of sieves with different sized perforations are used and the cleaning is a process of gradually sifting out smaller and smaller particles. It is not only the size of the perforations which determines the quality and quantity of the seed cleaned; other important factors include the precision of the perforations, the angle at which the sieves operate, the amplitude and speed of movement of the sieves, and the correct cleaning and maintenance of the equipment.

Sieves or screens may be made of flat perforated plate or wire mesh, and sometimes they may be three dimensional such as funnelshaped sieves. For small samples hand held sieves are adequate, but in larger scale cleaning a series of shaking screens is commonly used. In Brazil sieving with a mesh size of 32 per inch (approx. 12.5 per cm) was effective in separating seeds from chaff in Eucalyptus grandis. 84% of good seeds were retained and 89% of chaff was eliminated.

SORTING ACCORDING TO LENGTH

Cleaning with sieves relies on the separation of seeds, with diameter the critical factor. Fractionating according to length cannot be done with sieves, but this is possible with an indented cylinder. In addition to its use for separating good seed from the impurities, the equipment is used in agriculture for separating seed mixtures and can also be used for grading seeds.

The equipment consists of a slightly inclined horizontal rotating cylinder and a movable separating trough. The inside surface has small closely spaced hemispherical indentations. Small material is pressed into the indents by centrifugal force and can be removed. The larger material flows in the centre of the cylinder and is discharged by gravity. Depending on the type of impurities, the seed may be separated via the indentations or by passing down the cylinder.

BLOWING

Cleaning by blowing is a very important and widely used method. It is based on the principle that any object can float in an airstream of sufficient velocity. For separation in an airstream there are three possibilities: falling, floating or rising. The behaviour of the seed and other matter will depend on their weight, their resistance to the flow of air (volume and shape), and the velocity with which the air moves.

The operation of blowing is often referred to as winnowing or fanning. In its simplest form the uncleaned seed is thrown into the air on a windy day. The components separate out and the desired ones are retained. Indoors the airstream from a fan can be utilized.

Hand winnowing has been used successfully in Thailand to separate full seeds from empty seeds of Pinus kesiya (Bryndum 1975). The initial separation was followed by a second winnowing of the discarded fraction. Cutting tests showed that the original full seed percentage of 82% was raised to 98% in the improved fraction, while the discarded fraction (approximately 10% of the total volume) contained only 18% full seed or about 2% of the total full seed in the seed lot.

One operator took 8 minutes to winnow 1 kg. Loss of the very smallest and lightest full seed in the separation process is usually not serious since these seeds are likely to germinate slowly and to be low in vigour.

Simple winnowing machines can be easily constructed and Yim illustrates a machine made entirely of wooden components which is used in Korea.

Laboratory blowers can be classified as either the pneumatic type—a fan is situated near the air intake and pushes air through the system, or the aspirator type—a fan is situated near the air outlet and pulls air through the system, creating a partial vacuum. Small laboratory cleaners such as the 'Brabant' cleaner, and the 'Kamas' laboratory aspirator are available. Another seed blower in common use is the South Dakota Blower. The principle of blowing is that a sample of seed, when suspended in an upward stream of air of a certain velocity, will split into a light fraction and a heavy fraction, the light fraction being carried upward and the heavy fraction falling down. The two fractions are caught apart from each other. The heterogeneity of the heavy fraction can be further reduced by subjecting it to a second blowing at a higher wind velocity. In that way a light fraction, a medium fraction and a heavy fraction are obtained. The blowing apparatus in the South Dakota Blower consists essentially of a centrifugal blower, the outlet of which is connected to the bottom end of a vertical tube of a few centimeters' internal diameter and about 50 cm in length.

The sample is held on a fine wire gauze at the bottom of the tube. A built-in valve allows the wind velocity to be set at a rate that has been found optimal for the species. The lighter particles are blown up and trapped by baffles near the top of the tube, while the heavier are retained at the bottom.

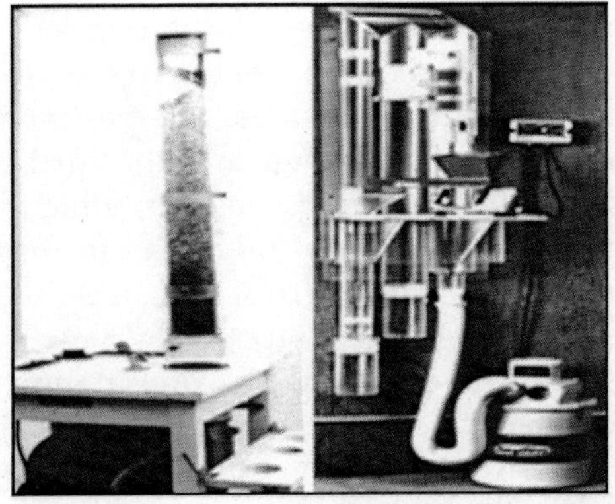

Fig. Electrically Operated Laboratory Seed Blowers: (A) South Dakota Blower (Division of Forest Research, CSIRO Canberra, Photography by Allan G. Edward). (B) Barnes Tree Seed Separator

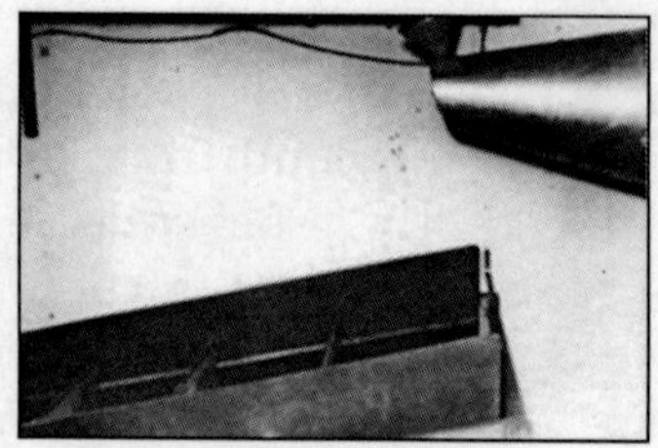

Fig. Locally Made Seed Cleaner used in Zimbabwe (A) Internal View of Conical Shield Showing Diffusing Baffles (B) In Operation, Showing the Compartmented Receptacle.

A more complex blower developed in Canada is described by Edwards. It consists of four plexiglas tubes of varying diameters which cause differential air velocity and, by choosing the appropriate combination of tubes, it is possible to use the equipment either to separate seed from chaff or full seeds from empty seeds. It works well with large seeds *e.g.* Abies amabilis but is not suitable for very small, light seeds such as Betula or Chamaecyparis.

In Zimbabwe a home-made cleaner was constructed by fitting a tapered aluminium sleeve with internal baffles onto a constant speed domestic fan (Seward 1980). The narrow end of the sleeve projects over a compartmented receptacle; full seeds fall into the nearest compartment, while the lighter impurities and empty seeds are blown into the further compartments. A similar locally made seed winnowing chamber, which uses a stream of air from an electric fan, has been successfully used in the Solomon Islands to separate debris and empty seeds from dry seed batches of Swietenia macrophylla and Campnosperma brevipetiolata.

Many seed cleaning machines use a combination of winnowing and screening. A coarse upper screen removes larger material, a lower fine screen stops the seeds and lets through fine matter, and then the seed fraction passes through a transverse or nearly vertical airstream from a fan to remove chaff and empty seeds. The air-screen cleaner is the basic equipment of seed cleaning plants. The size of air-screen cleaners varies from the small two-screen model to a modern precision model, which uses several top and bottom screens and in one operation as many as three air separations.

LIQUID FLOTATION

Cleaning by flotation relies on the principle that the density of the seed of a given species is specific both for filled and unfilled seed.

There are two basic methods used:

1. Density method in which liquids with a density or specific gravity between that of the full and empty seed are used. The specific gravity of the liquids used is usually below 1.0 and such that the full seed sinks and the empty seed and light debris float.
2. Absorption method in which water is used and, although both full

and empty seeds float initially, after some time the full seeds absorb water, become heavier and sink. The time of soaking can vary from a few minutes to several hours. This method is useful where there is a very small difference between the specific gravities of the full and empty seeds. The seed must be re-dried after being separated.

The flotation methods can separate out insect attacked, mechanically damaged, and immature seeds from filled mature seeds. The density method can only be applied if a liquid of suitable density is available which is not harmful to the seed.

The application and the problems encountered are discussed by Simak (1973); both this and a method for separating viable from non-viable full seeds which was developed by the same author, involving an element of pregermination, are described.

FRICTION CLEANING

Most debris can be removed from the seed by air-screen combinations, but leaf fragments, resin particles and other objects of similar size and density to the seed are difficult to remove.

Friction cleaning relies on the principle that any object falling or sliding over a surface encounters a certain friction. The movement of the particle is proportional to its weight and to a coefficient of friction which depends on the nature of the particle's surface and the surface on which it moves. Separation of seed from debris is made on an inclined cloth or rubber belt on the basis that the angle necessary for the run-off of the seed differs from the angle necessary for run-off of the debris. A continuous upward moving belt removes seeds downwards by gravity and the lighter debris upwards by friction.

SPECIFIC GRAVITY SEPARATION

This method makes use of a combination of weight and surface characteristics of the particles to be separated. It is a method which is finding increasing use in separating and grading tree seeds.

The specific gravity (SG) separator employs a flotation principle. A mixture of seeds is fed onto the lower end of a sloping perforated table. Air, forced up through the porous deck surface and the bed of seeds by a fan, stratifies the seeds in layers according to density, with the lightest seeds and particles of inert matter at the top and the heaviest at the bottom.

An oscillating movement of the table causes the seeds to move at different rates across the deck; the lightest seeds float down under gravity and are discharged at the lower end, whilst the heaviest ones are kicked up the slope by contact with the oscillating deck and are discharged at the upper end.

Deck coverings of linen, plastic, and wire mesh have been used to distribute the air uniformly beneath the seeds and provide the correct push to the heavier seed in contact with the deck. A closely woven covering such as linen gives the best results for small seeds. On modern gravity separators, it is possible to control the feed speed, the tilt of the table in two directions, the speed of oscillation and the force of air differentially at various points along the deck.

The combination of these different controls makes it possible to adapt the machine to handle a wide variety of species and seed lots.

The specific gravity separator will separate particles of the same density but of different size, and particles of the same size but of different densities. It will not separate efficiently particles which differ both in density and size, *i.e.* separate a larger but less dense particle from a smaller but denser particle. It has been found practical for cleaning the chaff from some eucalypt seeds and for grading pine seeds.

Purity% of uncleaned seed of E. grandis after extraction is about 10%. This was raised to 95% purity with 95% germination by SG separator treatment.

SEED GRADING METHOD

Within a single species there is a variation of seed dimensions due to environmental influences during the development of the seed and to normal genetic variability. The performance of the seed immediately after germination is related to seed size and, in order to produce a crop of seedlings which will emerge and grow evenly in the nursery, size grading of seeds can be a useful practice.

Size grading can also assist mechanical sowing of seeds. The practice should be used with caution in the case of seeds collected from seed orchards having a restricted number of clones. Since part of the variation in seed size and shape is genetic, grading of seed orchard seed could lead to excessive genetic differentiation and loss of genetic diversity within each of the fractions obtained by grading.

Fig. Air/Screen Seed Cleaner used in Humlebaek, Denmark

Fig. Damas Gravity Seed Separator

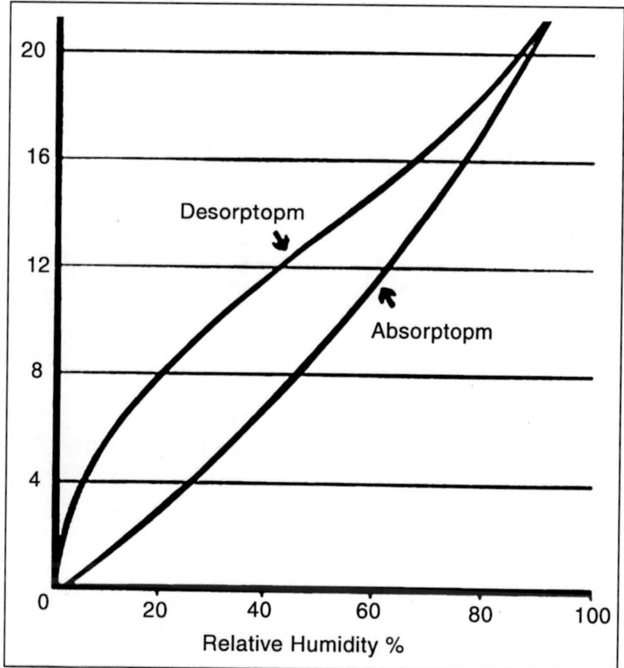

Fig. Equilibrium Moisture Content of Wheat Seed, Showing Separate Curves for Desorption and Absorption.

The methods of grading seed vary little from those used in the cleaning process. Sieving and screening, cylinders, air blowing, flotation and specific gravity separation can all be used effectively for size grading tree seeds.

Although grading itself is a relatively simple operation, seeds of certain species have to be properly processed before they can be graded, for example the spongy exocarp of teak fruits and the calyx tubes of dipterocarps must be removed in order to get the full benefits of grading.

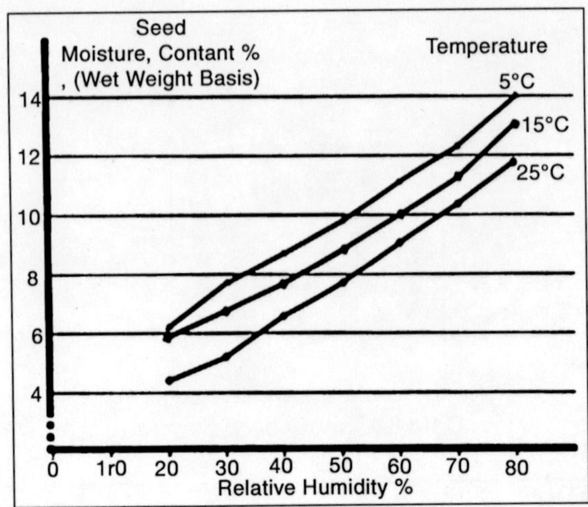

Fig. Moisture Content Percentages of Fresh Seed of *Pinus Palustris* in Equilibrium with Air at Various Temperatures and Relative Humidities

The cleaned stones of Gmelina arborea have been graded by using square-mesh screens of 7,9 and 11 mm mesh; germination varied from 84% for the smallest to 111% for the largest size class (there are usually 1 to 3 seeds per stone).

TYPES OF DRYING KILN

Turnbull (1975 c) classified kilns under the following types:

- Stationary tray kilns
- Vertical progressive kilns
- Horizontal progressive kilns
- Rotating drum kilns
- Portable kilns.

STATIONARY TRAY KILNS

The simplest kilns are convection kilns consisting of anything from a stove-heated room to a more complex structure. Basically they consist of a heating unit above which is an extraction chamber. Heat enters the chamber at the base and passes up through tiers of trays containing the fruits or cones. Since the air becomes cooler and moister as it rises through the cones, the efficiency of extraction is not uniform in all parts of the kiln. Some trays must often be moved to different levels and re-treated.

Such kilns are relatively cheap and they require no special technical skills to operate. They have the disadvantages that they are relatively inefficient, fire risk is often high, and temperature and humidity control are rarely adequate. They can be improved to some extent by the provision of forced ventilation. In Zimbabwe most cones are air-dried, but this is supplemented by a simple kiln using tobacco-curing flues. Because there is no mains electricity in the area air circulation is by convection and the kiln is wood-fueled. Safe maximum temperatures are 48°C for Pinus elliottii, P. taeda and P. kesiya and 60°C for P. patula. The kiln holds 36 trays, equivalent to 9 bags (about 22 bushels or 8 hectolitres) of cones, which will open in one shift of 8 hours.

Fig. Solar Drying of Pine Cones in Thailand, Showing Polythene Lids

Fig. Kiln Seasoning. Stacked Trays of *Pinus Radiata* Cones Entering a Kiln in New Zealand

Fig. Interior View of a Kiln

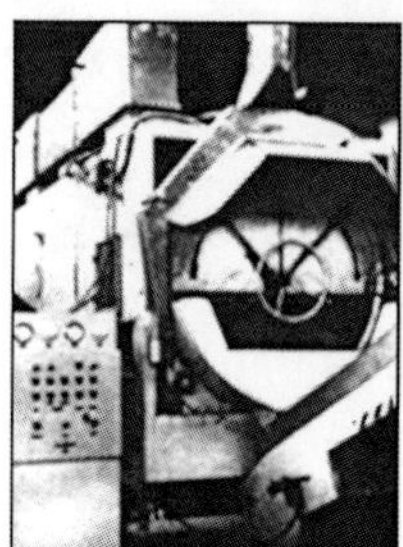

Fig. Rotating Kiln with Discharge End Open to Show Drum; Control Panel.

In Honduras cones of Pinus oocarpa and P. caribaea, after a period of precuring, are successfully dried in a forced draught hot air kiln modified from a solar kiln for drying timber. Ventilation is provided by two large diameter fans powered by a one horse-power electric motor, while heating is provided by flue radiator and a furnace using waste cones and firewood. Air can be recirculated if required by opening a valve which connects the air entry and exit passages. The kiln holds 32 hectolitres of closed cones, arranged in 8 stacks of 8 trays each, with 50 litres of cones per tray. The temperature regime is 4 hours at 40°C followed by 10–14 hours at 45°C for P. caribaea, while the thicker cones of P. oocarpa receive a constant 50°C. Cone opening is usually complete in 12–18 hours. Seed yield is 160–250 g per 100 litres of closed cones for P. oocarpa, 125–625 g per 100 litres for P. caribaea.

Electric ovens, preferably of forced draught type, are useful for drying small lots of eucalypt capsules. In Brazil electric ovens, consisting of a large chamber with a series of sliding shelves and a capacity of 80 kg of material, are used for drying eucalypt capsules. Drying time is 24 – 36 hours at 45° C. A forced draught oven has been used sucessfully in the USA to dry pods of Prosopis. The pods are placed in the oven at 32° C for 18 hours. The dried pods are then placed in an electric scarifier for 10 – 15 seconds, after which the seeds are separated by removing light debris in an air-column blower and sieving out the pod fragments with a size 11 or 12 screen (mesh size 1.85 or 1.70 mm).

In Sabah pods of Acacia mangium are dried in a simple drying chamber which incorporates an electric heater and a domestic fan (Bowen and Eusebio 1981b). Pods are held in stacks of trays 0.7 × 0.7 m square, each with a wire mesh base.

Air entry temperature is 45°C. Pods are separated by colour and treated as follows:

- Black pods—no precuring, dried for 24 hours in drying chamber
- Brown pods—precure in shade for 48–72 hours, then dry in chamber for 24–48 hours
- Green but full-sized pods—precure in shade for 72–120 hours, dry in chamber for 48 hours. The same method is used for Albizzia falcataria.

In place of trays, steel-mesh cylindrical drums are sometimes used to hold the seed-bearing branch clippings of eucalypts. Kilns of this type are in use at more than one site in Tasmania. One kiln is approximately 9 m × 4 m and has an extraction capacity of about 4500 kg of seed per annum. It has insulated walls with a transparent roof, formed of two layers of acrylic sheeting with an air space between, which permits solar heating in summer. In winter a false reflective ceiling is put in to reduce loss of radiant heat. The kiln is maintained at a temperature of 40° C. After 36 hours the capsules have opened and the cylinders are rotated and shaken to dislodge the seeds. If no special equipment is available and only a few fruits or cones are to be dried, they may be placed on a warm radiator. They should never be left to open where it is too hot to rest the hand comfortably.

VERTICAL PROGRESSIVE KILNS

The cones are placed in the kiln on a set of trays, one above the other, in a tower-like structure; trays descend during cone treatment. From the bottom of the kiln the hot dry air hits the lowest tray, containing fairly dry cones; passing through the cones, the air loses some heat and takes some moisture from the cones. In the second tray from the bottom, cones have somewhat higher moisture content and the air is less hot and less dry. In the topmost tray, the cones, still having the original moisture content, are engulfed by warm, humid air.

At the right moment, a tray is taken away at the bottom, and the whole tray set descends one step, while a new tray with fresh cones is put on the top of the pile. Hot air can circulate simply by convection, but forced ventilation makes the operation quicker and more regular. The distance between two successive trays should be sufficient to contain open cones which have a volume 2 to 3 times that of closed cones. The time required to complete drying varies according to the initial moisture content of the cones, the volume of circulating air, and the type of air circulation (convection or forced ventilation).

In some kilns, cones are forced to fall down from one tray to the next below by opening the tray bottom, or by inclining the tray. In other types, trays are replaced by slow moving bands; cones move on a band till they fall to the one below. In the last two types, cones are shaken as they fall from tray to tray, so that most of the seeds come out: special devices under the bottom tray collect the seed, while cones pass to further handling.

HORIZONTAL PROGRESSIVE KILNS

The operation of this type of kiln can be illustrated by the Italian designed horizontal kiln described by Gradi (1973). A continuous conveyor belt of perforated steel plate on which the cones are spread passes through an insulated galvanized tunnel. By means of radiators and ventilators installed in chambers on one side of the horizontal belt, an air movement in the form of a spiral is obtained from the entrance to the exit of the drying tunnel. In the first few metres of travel the cones undergo a pre-drying stage at a temperature about 10°C lower than the final temperature. In successive sectors of the drying tunnel higher temperatures are used, usually in the range of 45°–50°C. The spiral movement of the air creates a turbulence which greatly accelerates the drying process.

The kiln described by Gradi is designed to process large quantities of cones. More versatile equipment has been produced to process and maintain the identity of small seed lots as well as large ones (Isaacs 1972). This comprises a long horizontal kiln having small bins with removable trays rather than a conveyor belt. The kiln is provided with portable heaters which are rotated each day when unloading the kiln to ensure that the hottest air is on the driest cones. The static pressure and air flow is controlled through each bin. This equipment has a particular advantage in being very easy to clean. An alternative simple method of preserving the identity of small seed lots is to use small nylon bags. They allow free air circulation and the same bag can be used for the cones from the time of their insertion in the forest throughout transport and extraction.

The advantages of the horizontal kiln over the vertical kiln are:

- The expensive process of elevating cones to the entrance of the conventional vertical drier is avoided;
- Any possibility of damage to the cones in falling from one level to the next is avoided, because they remain stationary on the conveyor during all phases of the drying process;
- The high productive capacity;
- The ease of installation, inspection and maintenance. The relatively high cost of the equipment requires a high through-put of cones for optimum economy.

ROTATING DRUM KILNS

This type of kiln can be found in many modern seed extraction plants. The basic operation is as follows:

Cones are put in a cylinder made of perforated steel plate, rotating on a central axis. The cylinder is contained in a box, where forced air circulation is provided by an electric fan. During the extraction the cylinder rotates continuously, shaking the cones. The air temperature increases gradually, from room temperature to the maximum fixed level. The strong ventilation, the progressive heating and the continuous shaking make the cones dry and open in a short time. Seeds released from cones pass through the holes of the plate, and are blown at once out of the kiln. Temperature and ventilation control is generally fully automatic.

Seed is taken out of the hot air as soon as it comes out of the cone, in order to avoid seed damage. The capacity of the cylinder is reduced so that extraction can be done separately for cones from different provenances, even if the quantity is small. Kilns of this type may have electrical heating or a separate heating unit which can use different fuels.

The cylinder, the electric motor, the ventilator and the equipment for air and temperature control are contained in a steel box, above which is a pre-drying box with two floors. From this box, cones fall into the cylinder, where they are exposed to gradually increasing temperatures (40–45–50–60°C) for three to four hours. At the end of the extraction, reversing makes the cylinder open, the open cones fall down and other cones come into the kiln.

Modern rotating drum kilns are generally of compact metal construction and of moderate size so that they can be placed in small, inexpensive buildings. Their output is relatively low and they are generally planned to work in a set of two or three depending on the requirements of the extractory. Their advantages include the relatively short drying time and the shaking of cones during drying which extracts the seed and makes a separate tumbling operation unnecessary. This type of kiln is widely used for modern seed extraction of conifers and is being used for the extraction of eucalypt seed in Australia.

Fig. Portable Cone Kiln

Fig. Cone Tumbler in Humlebaek

Fig. Double-Storied Tumbler Room Showing (A) Inclined Ramp to Upper Floor (B) Cowl at Discharge End of Tumbler.

PORTABLE KILNS

For small-scale extraction or to provide seeds for modest plantation programmes or for research, large commercial cone drying kilns are neither satisfactory nor economical. In research work it is often vital that seed be extracted from small lots of cones without the possibility of the seeds being mixed with other lots.

A kiln designed to dry small lots of pine cones is described by McConnell. It features economy, safety, portability, versatility and a small load size. The kiln consists of a heating mechanism based on bottled gas, a fan, a control panel, and both a drawer type chamber and a cabinet chamber to hold the cones. The capacity is 50 bushels (18 hectolitres or 1.8 m^3). The control panel is designed to sound an alarm or provide a visual warning of any malfunction in the equipment, and thermostats ensure that the temperature cannot, under any circumstances, exceed 74°C. The portable kiln described above could find wide application for small operations through its economical operation and advanced safety features.

SAFETY PRECAUTIONS

Any artificial heating involves a fire hazard and this is particularly true in handling cones, since dust, resin and dry cone scales are all highly inflammable. Stringent fire precautions including a ban on smoking should be enforced, fireproof, non-wooden construction materials should be used and arrangements made for frequent removal of inflammable dust and debris by vacuum equipment or other means.

Other precautions are necessary when drying certain dry fruits and seeds of certain species. Dust masks need to be worn when handling species such as Platanus spp. which release fine hairs which might be breathed into the lungs.

SEPARATION

When fruits and cones open after drying, some seeds fall out easily as a result of manual stirring, rotation in rotating drum kilns or, in certain vertical progressive kilns, from the shaking of the cones as they fall from one tray to another. But many seeds are left inside, especially in those drying techniques where the cones remain static. They must be removed as soon as possible after drying is complete.

In some species a thorough manual shaking is sufficient to extract the remaining seed. The capsules of eucalypts need to be vigorously shaken, particularly if not fully mature, because abscission of the seeds from the placenta may be only partially complete. Failure to shake slightly immature capsules properly can result in only the chaff being released. The fertile seeds are usually attached to the placenta near the bottom of the loculus so that, after dispersal of the chaff, immature capsules may, on superficial examination, appear to be empty.

Cones may be shaken in coarse sieves to release the seeds, but more vigorous treatments are needed for some species. Those most widely used are tumbling for conifers and threshing for hardwoods.

TUMBLING

A tumbler is a rectangular or round container or drum mounted horizontally on its long axis. As it turns, cones tumble about; interior baffles often accentuate the jarring and tumbling action. Seeds fall from open cones through the high-strength wire mesh which forms the sides of the tumbler and into a hopper or trays or onto a moving belt.

The tumbler can be operated by hand or mechanically driven, depending on the scale of the operation. Some drums may be closed at both ends and emptied and refilled at the end of each operation cycle (Morandini 1962). In more modern designs a continuous operation can be achieved with an inclined cylinder open at both ends. The cones are fed in at one end and during rotation roll slowly to the other end where they are discharged. Variable speeds and tilt are set for each species. The speed determines the rolling and pitching effect on the cones, while the tilt determines the length of time the cones remain in the tumbler (Turnbull 1975 c). Small types of tumbler are easily transportable. Fisher and Widmoyer describe a small tumbler of ½ bushel (18 litre) cone capacity made from a modified domestic washing machine.

In Zimbabwe a hand-operated drum tumbler 2.43 m long is used, fed from a chute from the floor above. By building the tumbling room on a slope, the cones can be easily transported into the upper floor (Seward 1980). The drum holds one bag of cones and it takes one minute to tumble one charge and to recharge the drum. The seeds drop through the 18 mm mesh of the drum into a collecting tray below and the empty cones are discharged into a trolley.

It is important that tumbling should be carried out as soon as possible after drying, because open cones exposed to cold wet air can reclose in a short time. If tumbling cannot follow immediately after drying, the opened cones should be stored in warm, dry conditions in the interval between these operations. The time required for tumbling depends on the species and the condition of the particular lot of cones being handled. Seed in some species such as Larix decidua and Picea abies may be held tightly in the cone and long periods of tumbling are sometimes required to extract them. Special machines, such as a large potato-peeling machine or a tumbler incorporating saw blades are effective on difficult species like these.

Alternatively, the cones may be rewetted and then redried to promote fuller opening of the cone scales. Haverbeke found that, after the first tumbling in Pinus sylvestris, soaking the cones in trays of water at 30°C for about half an hour until the cones softened and started to close, followed by thorough air-drying until the cone scales opened again, gave good results.

The yield of seeds from the second tumbling averaged 36% of that from the first. There was a marked difference between provenances, from 18% for a Scottish to 84% for a Spanish origin. The additional yield from the second tumbling fully justifies the extra cost for rare and valuable seeds such as those obtained from controlled pollination.

Mechanical damage can easily be inflicted on seeds if excessive tumbling speeds are used or if the tumbler is filled with too many cones. Speed of rotation and time of treatment should be adapted to the cone and seed characteristics of the species being handled. It is better to leave some seed in the cones than to spend money on extracting seeds most of which will be severely damaged in the process.

THRESHING

Extraction of seed from dry fruits of many hardwood species is accomplished by threshing. Seed extraction in species such as Cercis, Catalpa, Robinia and Liriodendron is easily accomplished by spreading the fruits on a platform, sometimes on a straw mat or other suitable material, and beating them with a flail or slender pole. For large quantities mechanical threshers used in agriculture can be adjusted for tree fruits by altering the distance between the crushers. In chile pods of Prosopis tamarugo are ground in a stone mill set at 4 mm and the seeds are then separated by sieving and floating the milled product.

A modified cereal huller has been used as a thresher for Prosopis pods in the USA. The machine, described in Ffolliot and Thames, will thresh 1 bushel (36 litres) of pods in 1 ½ hours; approximately 160 hours would be required to thresh the same quantity by hand. In the Philippines fruits which do not readily release their seeds are put in a sack and beaten. Separation is done by use of screens. For every seed size one screen is used with a mesh larger than the seeds, to separate them from fruit fragments and other large impurities, and another with a smaller mesh which retains the seeds and allows the fine impurities to pass through.

In Sabah seeds of Acacia mangium are separated from the pods, after drying, by rotating them for 10–15 minutes in a cement mixer together with blocks of hard timber 10 × 10 × 15 cm. A similar use is made of a cement mixer for pods of Albizzia falcatariabut, because separation of the seeds is easier in this species, it is not necessary to include the timber blocks.

Several types of mechanical thresher suitable for Acacia pods are described by Doran *et al.*; they include a hand type model of resilient tapered thresher, a rotating drum, a flailing thresher and a peg drum thresher. Many acacias give off a very irritating dust during threshing and protective equipment should be worn by operators. More robust methods, such as pounding the fruits with a wooden pestle or putting them through hammermills, must sometimes be applied. For some species special equipment has been developed such as the dehuller of Juglans. Hammermills consist of a hooded inlet or hopper, a central chamber containing a series of hammers which rotate about a central shaft, and removable outlet screens of different mesh. The outlet screen must have holes large enough to let seeds pass through without damage. Fruits are fed into the mill continuously during separation. Care must be taken to operate hammermills at comparatively low speeds, 250 – 800 revolutions per minute, to avoid injury to the seeds. The Dybvig separator works well on dry as well as on fleshy fruits.

CONTROL OF MOISTURE CONTENT

After seeds have been cleaned and graded, they are ready for sowing in the nursery. If, however, they are to be put into storage, it is necessary to check their moisture content (MC) and, if necessary, adjust it to the optimum level for storage of the species in question. Adequate facilities for testing MC should be available in the seed processing depot.

For orthodox seeds, which comprise most coniferous and many hardwood seeds, adjustment of MC if needed means further drying. This is described in the following section. Much less commonly and only in the case of recalcitrant seeds which must be stored at a high MC it may be necessary to moisten the seeds in order to raise MC to the optimum for storage.

For example promising results have been achieved for Acer pseudoplatanus by soaking the seeds in water for two or three days and immediately afterwards freezing and storing them at about -7°C in plastic sacks. Other genera, such as Quercus and Castanea, which have been lightly dried under cover in order to loosen their husks or involucral attachments, may benefit from soaking to restore MC to the optimum level (*e.g.* 40 – 45% for Quercus robur, Holmes and Buszewicz 1956, Suszka and Tylkowski 1980), before being placed in moist, cool storage.

RELATIONSHIP OF SEED MOISTURE CONTENT TO ATMOSPHERIC HUMIDITY

Seeds, like cones and fruits, are hygroscopic materials and, when detached from the parent tree, they lose or gain moisture to or from the surrounding atmosphere until their moisture content (MC) reaches a point of equilibrium with the humidity and temperature of the surrounding air. This is known as the equilibrium moisture content (EMC). Once it has been reached, it will be maintained as long as the humidity and temperature of the air remain constant; if they change, the seeds will again lose or gain moisture until a new EMC is reached. Wood is another good example of a hygroscopic material and behaves in a similar way.

"Wet" seed surrounded by "dry" air will lose moisture and therefore weight, while "dry" seed surrounded by "moist" air will gain them. In order to devise the most suitable methods of drying and storing seed, it is necessary to be able to quantify the moistness of both air and seed.

AIR HUMIDITY

Moisture in the atmosphere is in the form of water vapour, but air can hold only a limited quantity of water vapour. If this is exceeded, the air is said to be saturated and the excess moisture condenses as dew. The exact weight of water vapour (WV) which the air can hold at saturation depends on the temperature, as shown in the following table:

Temp. °C	-10	0	10	20	30	40	50	60
Weight of WV at Saturation (g WV Per Kg of Dry Air)	1.6	3.8	7.6	15	27	49	87	152
Density of Dry Air (Kg Per m^3) at a Pressureof 760 mm	1.34	1.29	1.25	1.20	1.16	1.13	1.09	1.06
Weight of WV at Saturation (g WV Per m^3 of Dry Air)	2.1	4.9	9.5	18	31	55	95	161

Most of the time the content of water vapour in the air is less than that at saturation. Relative humidity (RH) is defined as the ratio (usually expressed as a percentage) of the quantity of water vapour actually present in the atmosphere to the quantity which would saturate it at the same temperature; alternatively, as the actual vapour pressure in the air as a percentage of the saturation vapour pressure at the same temperature. For the seedsman relative humidity is the most important measure of atmospheric humidity because the equilibrium moisture content of seeds is most closely correlated with it.

For example the MC of seeds will be very nearly the same when in equilibrium with air at an RH of 50%, whether the air temperature is at 10°C (absolute humidity or weight of water vapour present = 7.6/2 = 3.8 g/kg dry air) or at 50°C (absolute humidity or weight of water vapour present = 87/2 = 43.5 g/kg dry air).

Though the absolute humidity of one is over ten times that of the other, the relative humidities are the same and it is relative humidity which has the greatest effect on the EMC of seeds. The importance of RH to the EMC of seeds and the dramatic effect of temperature changes on RH explain the importance of heat in the drying of many seeds. From the table above it can be seen that air with 3.8 g water vapour/kg at a temperature of 0°C would be saturated; at 100% RH it would be useless as a medium for drying seeds. But if the same air were heated to 30°C and provided no additional moisture were introduced from outside the system, its RH would be reduced to 14% and it would become a highly effective drying medium.

MOISTURE CONTENT OF SEEDS

The amount of moisture in seeds is usually expressed as a percentage of their weight. Methods of measuring MC are described.

Moisture content can be expressed in two ways:

1. The weight of water expressed as a percentage of the initial "wet-weight" or "fresh-weight" of the seeds (= dry matter + water) or
2. The weight of water expressed as a percentage of the final oven-dry weight of the seeds (= dry matter only).

One of the greatest difficulties in the understanding and application of published results on moisture content derives from the fact that, in the past, both the "wet-weight" and "dry-weight" methods have been used, often with no indication as to which of them were applied in a particular case. According to the ISTA Regulations seed moisture content should always be expressed on a wet-weight basis. For orientation both formulae are given here with a conversion table.

$$\text{Percentage of Moist Content, Dry Weight Basis} = \frac{\text{Weight of Water}}{\text{Weight of Dry Matter}} \times 100$$

$$\text{Percentage of Moist Content, Dry Weight Basis} = \frac{\text{Weight of Water}}{\text{Weight of Water} + \text{Weight of Dry Matter}} \times 100$$

Moisture Content,% of Dry Matter (Dry Basis)

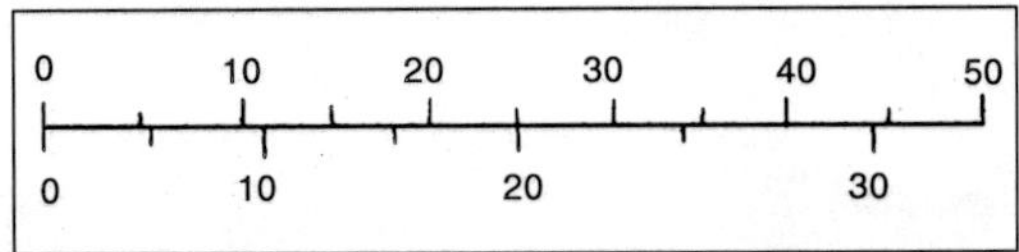

Fig. Moisture Content,% of Total Weight (Wet Basis)

Because of the limited amount of water vapour which it takes to saturate air, a relatively small quantity of seed can hold as much moisture as a great deal of air. One litre of seed dried from 50% to 9% MC (wet weight) at 30°C would lose about 450 gms of moisture to the surrounding atmosphere, enough to change the RH of about 15 m^3 (or 15,000 times its own volume) of air from 0 to 100%. In the case of sun-drying, the atmosphere is so vast that it can absorb this moisture without difficulty, but in an enclosed building the ambient air can quickly become saturated.

This explains why so much emphasis is put on adequate ventilation in kiln drying, in order to ensure that moist air is removed as it approaches saturation and is replaced by fresh, dry air. The same feature is an advantage when storing dry seeds in sealed containers. Provided that the seeds are correctly dried and the containers properly sealed, a relatively small volume of seed will come into equilibrium with a much larger volume of enclosed moist air without increasing its own MC significantly. If one litre of seed of oven-dry specific gravity 0.5, dried to 9% MC (wet weight), were to be enclosed in a 10 litre sealed container with 9 litres of moist air at 100% RH at 20°C, the total moisture content of the air would be only: $\frac{9 \times 18}{1000} = 0.16\text{g}$ Even if the seeds were to absorb all this moisture, it would only raise their MC from 50 to 50.16 g or MC% from 9.09 to 9.12%. The common prescription to fill sealed containers as full as possible with seed is a sound one, but it is based on the deleterious effects on many species of the oxygen in the enclosed atmosphere, not of the water vapour.

OTHER FACTORS AFFECTING EMC

Although relative humidity is the most important single factor affecting the equilibrium moisture content of seeds, it is not the only one.

Temperature

As already explained, temperature has a large indirect effect on EMC because, if absolute humidity is kept constant, relative humidity is directly related to temperature. It has an additional effect because EMC varies slightly with temperature even when relative humidity remains constant. The effect varies with species but very few data on forest trees have been published. An example quoted for an agricultural crop, sorghum, by Justice and Bass shows that at 50% RH the EMC varies from 12% at 49°C to 14% at -1°C. The difference is slightly greater in some other crops, but in all cases EMC decreases with increasing temperature and constant RH (even though absolute air humidity increases with temperature at the same RH).

Absorption and Desorption

For any species there is a difference of 1 – 2% in the EMC according to whether a moist seed is losing moisture to a drier atmosphere (desorption) or a dry seed is gaining moisture from a moister atmosphere (absorption). The EMC is always higher on desorption and it is the desorption curve which is important in the common situation of drying orthodox seeds from a higher to a lower MC for storage.

Variation in EMC According to Species

The MC of seeds in equilibrium with a given RH and temperature varies with species. The EMC for each species must be determined by trial. An important component in interspecific variation is the percentage of oil content in the seeds.

Seeds which store most of their food reserves as proteins or starch have a higher EMC at a given RH than seeds which store food as fats and oils, because the former are relatively hydrophilic, the latter hydrophobic. Among agricultural seeds wheat, with a low oil content of 2% and an EMC at 45% RH and 25°C of 10.4%, may be compared with Brassica oleracea, with a high oil content of 35% and an EMC under the same conditions of 6.0%.

Equilibrium Moisture Contents for 3 Orthodox Species

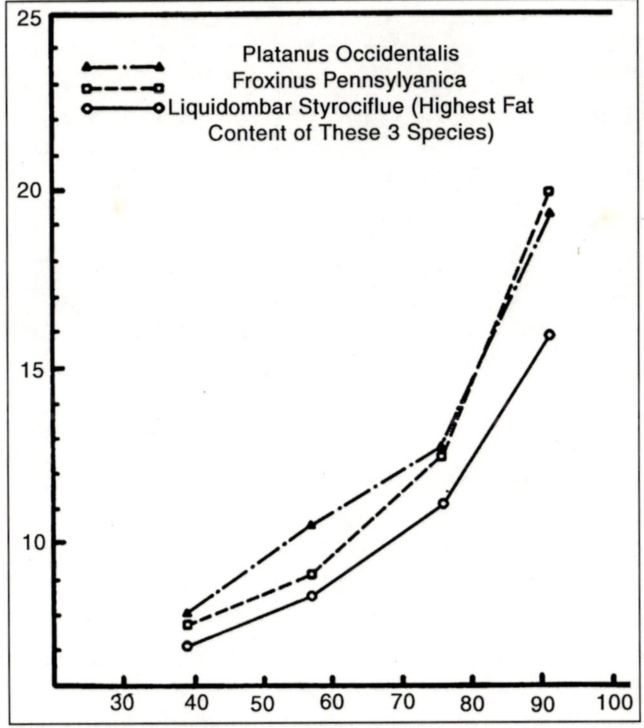

Fig. Equilibrium Moisture Contents for 3 Orthodox Species.

Equilibrium Moisture Contents for 4 Recalcitrant Species

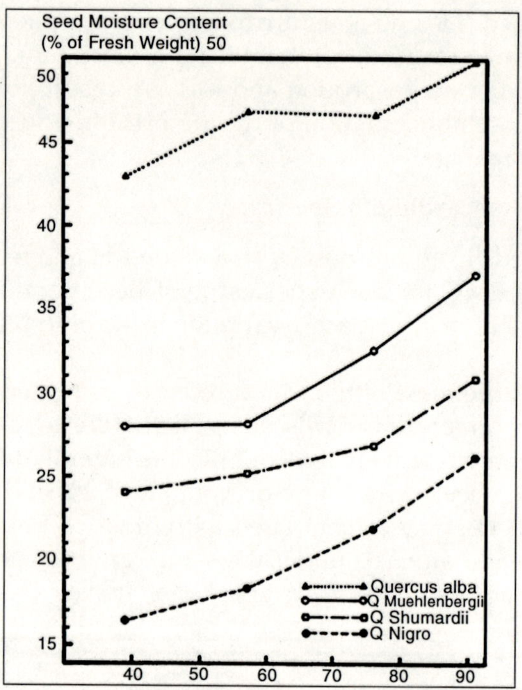

Fig. Equilibrium Moisture Contents for 4 Recalcitrant Species.

Detailed information on the EMC of tree species is sparse, and lacking entirely for tropical species. Some selected examples are given in the following table and in the attached graphs provided by F.T. Bonner.

Table. EMC% (Wet Weight Basis)

Species	Temperature	RH 10	20	30	40	40–55	50	60	70	95
Fraxinus sp.	not given	4.1	6.0	7.4	8.8	–	10.3	12.0	13.9	–
Picea abies	not given	2.4	4.2	5.5	6.7	–	7.8	9.0	10.4	–
Pinus taeda	4 – 5 °C	–	–	–	–	10	–	–	–	17

The graphs show EMC for four recalcitrant (Quercus) species and three orthodox broadleaved species. It should be noted that the EMC among the Quercus spp. is correlated positively with carbohydrate content and negatively with fat content.

Q. alba has the highest EMC, the highest carbohydrate content and the lowest fat content, followed by Q. muehlenbergii, Q. shumardii and Q. nigra. Similarly, Liquidambar has the lowest EMC, the lowest carbohydrate content and the highest fat content of the three orthodox species.

The same process of tissues reaching an MC in equilibrium with the ambient RH occurs during the drying of fruits for seed extraction as in the drying of the seeds themselves for storage. Usually the exact EMC is less critical in drying fruits, since shrinkage, splitting or scale-opening take place over a range of MC and the process is only carried on to the stage when the fruits release their contained seeds.

DRYING OF ORTHODOX SEEDS

For medium- or long-term storage of many species, a moisture content of 4–8% is recommended. This is considerably less than the MC of freshly collected seeds. Reduction of MC can be achieved in most species by placing the seeds in an ambient atmosphere of relative humidity (RH) 15–20% for a period sufficiently long to allow the seeds to reach an MC in equilibrium with the RH.

The effectiveness of air-drying depends on local climatic conditions. Reduction to an MC of 12–18% is frequently possible, provided that attention is paid to adequate aeration of the seed. Reduction to less than 8% is impossible in most temperate situations and in some areas of the wet tropics, because average RH remains too high. Thus RH in moist tropical West Africa is commonly over 80% in the wet season and over 70% in the "dry" season. An MC of less than 8% is unlikely to be achieved in these conditions. In areas where insolation is high, many species can be successfully dried to 6–8% MC by exposing them to direct sunlight, because the seeds and surrounding microclimate heat up and RH is consequently reduced. Also, for a constant RH, the equilibrium MC of seeds decreases with increasing temperature. Care must be taken to ensure that the seeds are as dry as possible prior to exposure and that they are moved frequently. In Honduras this method works well with Pinus spp. but is not recommended for Cordia because seed of this genus dries too quickly and can reach 4% MC, with damage to the tissues.

As pointed out by Harrington, artificial drying can be accomplished in two ways. One method is to raise the temperature of the air which, provided that no additional water vapour is introduced from outside the system, automatically reduces the RH. The other is to remove moisture from the air without changing the temperature, which also reduces RH. He quotes the example of air at 5°C and 90% RH which is heated to 35°C. Its RH is thus reduced to 15% and it can be blown through the seed by forced ventilation until the latter reaches equilibrium MC. On the other hand, air at 30°C and 90% RH, typical of moist tropical areas in the rainy season, would still have an RH of 40% even if heated to 45°C.

High temperatures can be extremely injurious to seeds, especially those which have a high MC. Generally, drying temperatures should not exceed 40°C and recent trends have been towards lower temperatures and greater air-flows to ensure safety in drying. Barner and CATIE recommend a temperature not exceeding 30°C in the early stages. One possibility is to dry seed in two stages, the first to about 11% MC using a temperature below 40°C, the second to about 5% using a temperature of 60°C. Provided that the MC is reduced to about 11% in the first stage, a second stage temperature of 60°C is considered safe for most agricultural species. On the other hand, there is some evidence that a high drying temperature, which does not affect immediate germination, may still affect subsequent longevity. For long-term storage for genetic resources purposes, a combination of low RH and low temperature (15% RH and 15° C) is recommended.

Where climatic conditions make it impossible to achieve a low enough RH by heating the air, it is necessary to reduce it by removing water vapour without raising the temperature.

This can be done by either:

- Refrigerating the air to below the dew point, condensing the water vapour on the cooling coils and then reheating the air to 35°C or
- Blowing air through a chemical desiccant, which removes the water vapour, and then through the seed.

Various desiccants are available *e.g.* silica gel, CaO, H_2So_4, Lithium chloride or anhydrous $CaCl_2$ but the most indestructible and easiest to reuse is silica gel. A good example of one drying technique is that furnished by the seed bank of the Regional Genetic Resources Project at Turrialba, Costa Rica. Because of the permanently high air humidity and high day temperatures there, hot air drying is not possible without damage to the seeds. Therefore a silica-gel type dryer, which maintains an RH of less than 15% at 25°C, is used, situated outside the drying room and ducted into it.

Small lots of seed that have first been air-dried to an MC below 20% may be placed in a sealed container with an equal quantity of silica gel that has been freshly dried at 175°C and cooled. The silica gel, seed and enclosed air will come to an equilibrium suitable for storage. See also the section on pp. 156–157 on "Use of dessicants in containers". The period of time needed for forced-draught air circulation to dry seeds to equilibrium moisture content is partly dependent on the accessibility of individual seeds to the air current. Seeds should be spread in thin layers on trays, with space for air circulation between the trays.

The MC of coniferous seeds, which have been extracted from the cones during kiln-drying, may already be close to that recommended for storage. But moisture may be reabsorbed during cleaning and dewinging, and in some species it may be added deliberately to reduce damage during these operations. Therefore MC must be tested and, if necessary, further reduced immediately before storage. At the same time, if cleaning and dewinging are carried out in warm, well-ventilated rooms, this will decrease the amount of drying needed subsequently.

SEED-DRYING METHODS

Drying seeds to the desired moisture level for storage may take a few days to several weeks, depending upon the species, the atmospheric humidity, and the equipment you use. The faster you dry the seeds, the less likely they will be to succumb to pathogens. The lower the humidity of the air in which the seeds are placed, the faster the seeds will dry. Seeds dry quickly at first, then more slowly as their moisture content nears that of the air around them.

PACKAGING SEEDS FOR DRYING

Small coin envelopes work very well for holding seeds while they dry, better than standard paper letter envelopes, which often have small holes at their corners through which tiny seeds can spill out. Another option is to wrap small seeds tightly in pieces of paper towel and secure the seams with tape. Whether you use envelopes or paper towels, package only a few seeds together.

If you have a lot of seeds, you'll have to use a lot of envelopes. Be sure you carefully label each envelope or paper towel with the date and the crop or species name (and the variety name if there is one). Note that although plastic bags and glass jars work well for storing dried seeds, they are not good container choices for seeds during the drying process.

DEHUMIDIFIER DRYING

If the air in your home is damp, as is often the case in coastal states and in the South, set up your seeds in a small room with a dehumidifier. You can use an inexpensive hygrometer and a thermometer to estimate how dry seeds are by tracking the relative humidity and maximum and minimum temperatures in the drying room. Measure the relative humidity and temperature daily.

Leave the seeds in the drying area for at least a week to be certain their moisture has come into equilibrium with that in the air. Ideally, follow this rule of thumb: The sum of the relative humidity and the storage temperature in degrees Fahrenheit must not exceed 100, as long as the temperature is less than 50 °F (10 °C). For example, if the relative humidity in your drying room is 40 per cent and the temperature is 40 °F (4 °C), then the sum is 80 (40 + 40), which is less than 100, so your seeds should dry just fine.

Most orthodox seeds will dry to an ideal 6 to 8 per cent moisture if they're held at 38 to 40 °F (3.5 to 4.5 °C) and 30 to 35 per cent relative humidity. Seeds of recalcitrant species like those of the oaks will dry to their appropriate moisture content if held at 38 to 40 °F (3.5 to 4 °C) and about 40 per cent relative humidity. As a homeowner, you don't have the wherewithal to purchase expensive moisture meters to be absolutely sure of having perfectly dried seeds, but these schemes will get you into the ballpark. You can easily see that keeping the relative humidity below about 40 per cent will reduce the seed moisture to below about 8 per cent, although more exact percentages depend upon species and seed lot. The table below gives some specific examples of the expected moisture content of vegetable-crop seeds when dried at 40°F (4.5 °C) at 45 per cent relative humidity.

If you live in a fairly humid area and have no dehumidifier, you can try drying seeds in your oven instead. This is easy, works well, and is safer for the seeds than is drying them in direct sunlight. Note that this method is primarily for larger seeds such as those of squash and beans. Tiny seeds such as those of cabbage and carrot seeds should dry sufficiently without heating in an oven. To prepare seeds for oven drying, do not put the seeds into envelopes or folded paper towels. Instead, spread the seeds in a thin layer on a cookie sheet and place the cookie sheet in the oven for 24 hours at 100°F (38 °C). Stir the seeds once or twice during this time to be sure all sides are exposed to drying air, and that should do it. Once the seeds are dry, you can package them in containers for storage. Some pinecones may require a higher temperature, 130°F (55 °C) or more; we note exceptions like this in the plant entries in part 2.

USING SILICA GEL

Following a dry-air treatment with a silica-gel treatment provides extra insurance that your seeds will be dried well enough. This is advisable if you live in the mid-Atlantic or southern states, or in other areas with very high relative humidity, or in general if you want to be absolutely sure your seeds are ready for storage. Silica gel is widely available at photosupply stores or by mail order from some garden seed companies and is very easy to use. The best bead size for drying seeds is 1/2- to 1/8-inch diameter.

To set up seeds to dry with silica gel, first measure out an amount of the deep blue silica gel equal to the weight of the seeds. Place either the silica gel or the seeds in a porous bag, then place both in an enclosed space such as a glass jar with an airtight seal. Make sure the silica gel is not touching the seeds. Place the containers in a room held at 59°F (15°C) for drying. You'll need to replace the used silica gel with fresh, either daily or when the colour turns from deep blue to pale blue or pink (which indicates that the gel has absorbed its fill of moisture).

Allow the seeds to sit, changing the silica gel as needed, for a few weeks to be sure they have dried properly. Larger seed lots and bigger seeds require longer drying times.

Some folks use powdered milk as an alternative drying agent to silica gel, but it is perhaps one-tenth as effective. Use it as you would the gel. Powdered milk can't be redried very well, so dispose of it.

REUSING SILICA GEL

You can reuse silica gel over and over again as long as you dry it out between each use. To do so, heat used silica gel in a warm oven to drive off the moisture. Set the oven at a temperature above 212 °F (100 °C) but below 275 °F (135 °C). Place the gel in a thick-walled Pyrex dish in a layer no more than an inch deep. Stir the gel a few times during the drying process; 1 quart (1.9 pounds) of gel should take about 2 hours to dry. You'll know it's dry when it has turned deep blue again. Alternatively, heat the gel in a microwave oven at a medium or medium-high setting for 3 to 5 minutes. If it is still not dry at the end of the cycle, stir it and heat it again for the same period. It may take about 10 minutes to dry a pound. Store the dried gel in an airtight container for future use.

Conservation of crop genetic resources is important to the long-term health of the world's food production systems. Genetic diversity provides the raw materials for selecting and improving plant traits such as resistance to pests, diseases, and environmental stresses. Genetic engineering has greatly increased our ability to manipulate genes for the benefit of agriculture, including transferring genes between unlike species. However, until individual genes that code for a specific trait or set of traits can be designed and developed in the lab, researchers and plant breeders must rely on existing genes. Thus, saving and preserving seeds, and their genes, is critical to future food security and stability.

Providing access to a reservoir of plant genetic resources has been the goal of ex situ (outside of the place of origin or natural occurrence) conservation at both the national and international levels. Ex situ strategies include the storage of plant genetic resources in seed banks, clonal repositories, and living collections. An extensive system for the ex situ conservation of plant genetic resources has been developed in the United States under the auspices of the Agricultural Research Service (ARS), the main research agency of the USDA (United States Department of Agriculture). The National Plant Germplasm System (NPGS), established to preserve and promote the use of plant genetic diversity, is a collaborative effort between state, federal, and private entities to acquire and manage plant genetic resources, including wild and weedy crop relatives, landraces, obsolete cultivars, and elite lines orpopulations of agricultural, horticultural, industrial, and medicinal crops. Though the NPGS focuses on building a strong, competitive U.S. agricultural industry, all germplasm held in the NPGS collections is made available to researchers around the world upon request.

Germplasm from all over the world is preserved in the NPGS system. Because many of the commercial crops produced in the United States are from nonnative sources, American agricultural productivity has depended on plant introductions from other countries, particularly from the tropics and subtropics. There are over four hundred thousandaccessions from more than ten thousand species in the U.S. germplasm reserves. Responsibility for maintaining and distributing this large collection is divided between different NPGS sites, such as the eight National Germplasm Repositories, four Regional Plant Introduction Stations, the National Seed Storage Laboratory (NSSL), and other NPGS sites. Each site is charged with maintaining different species.

The NSSL preserves the base collection of the NPGS and conducts research to develop new technologies for preserving seed and other types of plant germplasm. Seeds are stored either in conventional storage at –18 °C or in cryogenic storage (liquid nitrogen) at –196 °C. The National Germplasm Repositories are responsible for acquiring, preserving, increasing, evaluating, documenting, and distributing plant genetic resources of specific genera. United States germplasm collections of maize (corn), pumpkins, sunflowers, melons, cucumbers, and carrots are maintained at the North Central Regional Plant Introduction Station located in Ames, Iowa. The Western Regional Plant Introduction Station in Pullman, Washington, maintains lettuce, beet, bean, chickpea, forage and turf grass, and pea germplasm. Collections of pears, strawberries, blueberries, raspberries, and others are maintained at the National Clonal Germplasm Repository in Corvallis, Oregon, as living plants or, in the case of wild species, as seeds. Fruit and nut tree germplasm is maintained at the National Clonal Germplasm Repository in Davis, California.

Long-term storage of seed samples carries with it some inherent problems. The primary objective of seed banks is to maintain the genetic diversity and integrity of germplasm. Maintaining seeds in frozen storage requires adequate temperature and humidity controls. Even under ideal conditions, however, seeds eventually begin to lose viability—the ability to germinate.

Thus, periodically, new seed needs to be produced in order to replace aging seed samples. This process is referred to as regeneration. The loss of unique *genotypes* within a collection—whether from natural causes, small sample sizes, or random drift—nonetheless would constitute a change in the presence or frequency of genes within a germplasm collection.

During regeneration, plants are exposed to the risks inherent with agriculture—insects, diseases, drought, hail, temperature extremes, and wind, depending on whether they are grown under field or greenhouse conditions. When susceptible genotypes succumb to insect or disease pressures, the genetic variability and integrity of the collection may be compromised unless the genes lost are present in surviving genotypes. Thus, the need to regenerate a collection must be weighed against the need to minimize risks associated with regeneration. Additionally, genetic contamination through cross pollination or accidental mixing during post-harvest processes such as cleaning can also result in a loss of integrity.

Many small, independent organizations are also involved in seed saving. Together, they have created what is known as the heirloom seed movement. These organizations, groups, and individuals have helped bring about global awareness of genetic erosion—a reduction in the number of varieties, and, hence, genetic diversity—in commercially available vegetable and crop seed. Of the approximately five thousand heirloom varieties of vegetables available in the 1984 seed catalogs, 88 per cent were no longer available by 1998. On average, there is a 6 per cent loss in available varieties every year. Comparison of a USDA inventory of varieties available at the beginning of the twentieth century with a list of holdings in the NSSL at the end of the twentieth century revealed that only 3 per cent of this germplasm survived in American germplasm reserves. More than twenty-five small seed companies in the United States focus their efforts on slowing or preventing the loss of open-pollinated, heirloom seeds. Founded in 1975, Seed Savers Exchange collects, maintains, and distributes precious heirloom seeds through a network of eight thousand members. The vast collection of rare, heirloom seeds includes over eighteen thousand varieties of tomatoes, beans, peppers, squash, peas, lettuce, corn, melons, garlic, and watermelons from countries around the world.

Draped in a worn-out light green saree, she is sitting on the mud porch of her house, surrounded by little heaps of seeds and bamboo baskets. In her hands is a bamboo thresher which bears the appearance of black leather : it has been coated with cashew nut oil and cowdung so as to be more lasting. The door posts behind her are richly decorated in yellow, red, black, blue and green colours, in accordance with the Indian habit of making living spaces, however modest, aesthetically pleasing.

We are in Kamalpally, a small village of Andhra Pradesh, in Medak district, which lies in the semi-arid area of the Deccan Plateau. Agriculture is primarily rainfed here. A majority of farmers - both men and women—only have access to small plots of land on which they grow millets (sorghum, foxtail millet, little millet, pearl millet)and pulses (pigonpea, chickpea...). This dryland farming system is well adapted to the conditions, and some crops, like sorghum, can be grown both in the rainy season and in the post-rainy season.

This is the month of March, which corresponds to the end of the rabi season. The harvesting of sorghum and pearl millet has taken place, and the seeds are drying out in the sun in front of the house, forming large colourful halos on the ground. A young woman is sweeping the seeds back and forth with a bunch of leaves used as a broom. These leaves are from virungo negu, a plant known for its insecticidal properties, which will help the seeds keep better.

Once the seeds are dry the threshing work begins. The task, which demands hours, is entirely carried out by women.

Again and again, with a precise and refined gesture of the wrist, she throws the seeds in the air, blowing on them softly to remove small pieces of straw mixed in with the seeds. The sound produced as the seeds land on the bamboo surface reminds one of the music which emanates from a rain stick. The regular and precise rhythm of this rolling sound inhabits the villages after harvesting season. After a few throws, the pieces of straw begin dissociating themselves from the seeds : they all gather close to the edge of the thresher. She gently removes them, and she resumes the threshing. Now her gesture is slightly different, the pace is slightly faster. The seeds land diagonally on the surface. Her sharp and skilled glance stares into the seeds, as if to discern the good ones from the bad ones. And indeed, within instants, a pattern begins to emerge on the thresher : all the full and healthy seeds are collected at the bottom while at the top, near the edge of the thresher, a number of small, shrunken or broken seeds have gathered. She delicately pushes them over the edge and begins a new round of throws.

What seems to represent an infinity of threshing techniques are put to practice by the women to remove the stones and straw bits from the seeds, to separate the broken bits from the whole seeds, the small and shrivelled seeds from the healthier ones, the coarser seeds from the finest ones.

The seeds that have been selected for sowing are handled separately. They have been gathered in the fields before harvest, according to a variety of criteria (size of panicles, pest resistance, height of the stem...). For preservation purposes, they are mixed with some ash, and then deposited inside bamboo baskets smeared with cowdung. The top will be covered with fresh mud, transforming into a hard cover when it dries. Seeds are also preserved in earthen pots or aluminum tins until the following season. The grain used for consumption is stored in large vessels kept in the corner of a room. A handful of neem leaves used as an insect-repellent ensure that the grain remains free of pests during the months of storage.

Tree improvement programmes and the artificial regeneration efforts that accompany them require good seed supplies. Good seed must be collected and cared for properly to prevent loss of quality. Seed storage for short-term preservation of germ plasm is an integral part of most tree improvement programmes. This paper presents recommendations for the care and storage of hardwood tree seeds. For more detailed reviews of seed storage, readers should refer to Barton (1961) and Roberts (1972). Holmes and Buszewicz (1958), Sohönborn (1964), Stein *et al.* (1974) and Wang (1974) are excellent sources that deal with the storage of tree seeds.

PREDICTION OF DRYING PERIOD

If there is no previous experience of drying seeds of a particular species it may be necessary for you to do some experimental work to predict the approximate drying period. Seeds dry at an exponential rate until the equilibrium moisture content is reached. The rate of drying of different seed lots of the same species is fairly constant for seeds dried under the same environmental conditions. The length of the drying period can be predicted in one of two ways:

PREDICTION OF THE CORRECT DRYING PERIOD BY WEIGHT LOSS

1. Predict the current percentage moisture content and the percentage moisture content required for storage.
2. Weigh the seed sample.
3. Use these three values to calculate the weight of seeds at the required moisture content by using the following formula:

$$\text{Final seed weight} = \text{Initial seed weight} \times \frac{(100 - \text{Intial\% moisture content})}{(100 - \text{Final\% moisture content})}$$

4. Weigh the seed sample at regular intervals during the drying period until the weight of the seeds has reached this calculated value.

Notes and Examples

The prediction of the drying period will be more accurate if the initial moisture content is determined. However, this is a waste of seeds when moisture content can be adequately predicted.

1000 g of seeds with 12% moisture content were dried to 5% moisture content. What would be the weight of these seeds after drying? Substitute the values in the equation on the facing page:

$$\text{Final seed weight} = 1000 \times \frac{(100-12)}{100-5}\text{g} = 926.3\text{g}$$

Therefore when the initial 1000 g of seeds have been dried to 926.3 g, their moisture content will have decreased from 12% to 5%.

Equipment

- Coarse balance

PREDICTION OF DRYING PERIOD FROM MEAN DRYING CURVES

1. Do not waste valuable seeds for this, but use either excess seeds or those which are being discarded because they have lost viability.
2. Take at least two lots of seeds of the species and place to dry, using the method that you will use in practice.
3. Remove a sample of seeds and do an accurate determination of moisture content for each seed lot as described in Section III.
4. The mean of the two tests can be used as a guide because other seed lots of the same species should dry at a similar rate.
5. Repeat the determination daily and plot a graph of the drying curve (mean percentage moisture content against time) for that species under these drying conditions.
6. The work can be repeated with seeds of all species which are of interest and their drying curves plotted for different conditions.

Notes and Examples

If the seeds are very small and are dried quickly, the moisture content can be determined more frequently during the drying period to plot a more accurate graph.

Equipment

- Grinder
- Heat resistant dishes with covers
- Analytical balance
- Forced draught oven

- Desiccator
- Silica gel
- Tongs and oven cloth

USING DRYING CURVES

1. Use the graph that you have prepared for seeds of a particular species being dried under those conditions.
2. Predict the current percentage moisture content of the accession by using the table of equilibrium moisture contents. Select the final percentage moisture content that is required for storage of this species.
3. Find the value for each of the percentage moisture contents on the vertical 'Y' axis, follow this across to the curve and read off the day on the horizontal 'X' axis for each of the moisture contents. The difference between these two values is the drying period in days.

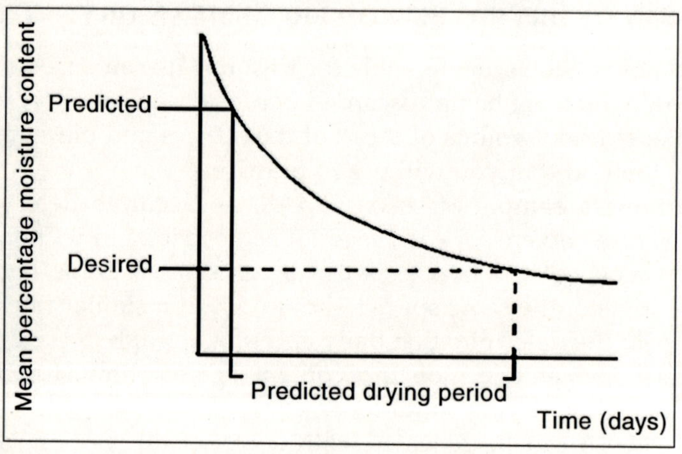

Fig. Typical Drying Curve

Notes and Examples

Example of how to predict the drying period by using a typical drying curve for small seeds of crops such as onion or cabbage, dried quickly under low relative humidity conditions:

Predict the moisture content by using the table of equilibrium moisture contents for your genebank which you have prepared. As an example take this as 12% and the desired moisture content after drying as 5%. Using the graph, the intersects from the curve to the time axis show the values 10 and 2.25 days.

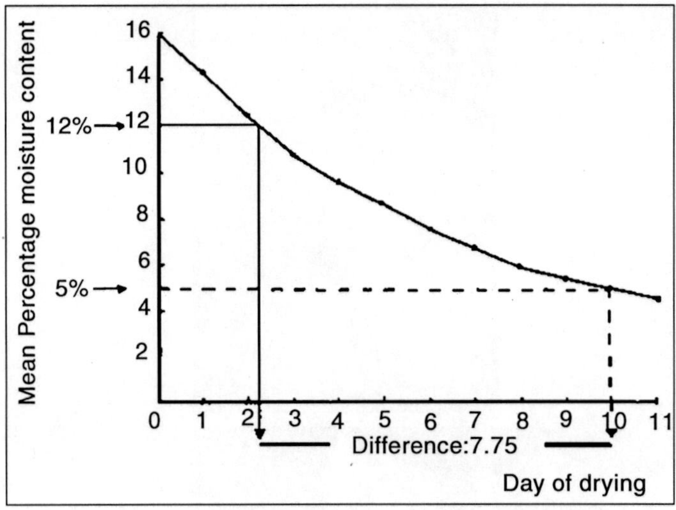

Fig. Typical Drying Curve of Small Seeds

Therefore, the time to dry the seeds to 5% moisture content under the same conditions will be the difference between 10 and 2.25 days. This can be approximated to the nearest day, in this case 8 days.

In practice it is unlikely that using the graph will give you a whole number of days, therefore you should approximate to the nearest day.

MIXING BEFORE STORAGE

When a large seed lot is to be stored in several containers, it is desirable that seed homogeneity between containers be maintained as far as possible, so that each container is equally representative of the seed lot as a whole. If seeds have been size-graded, mixing is carried out on the ultimate seed lots separated by the grading operation.

Thus, a given seed collection may be graded into a "large seed" fraction and a "small seed" fraction. Each fraction becomes a separate seed lot with its own seed lot number and separate mixing within each of these two seed lots may be done before storage to ensure homogeneity between containers.

METHODS OF EXTRACTION OF SEED

Seed extraction of some species is difficult, even after standard treatments of drying and tumbling or threshing. In the Philippines indehiscent hard pods of leguminous species such as Delonix regia, Pithecellobium saman, Cassia fistula, C. javanica and Parkia javanica have to be opened with a machete or knife and the seeds picked out individually. The pods of P. saman are sweet and relished by termites; if they are piled in a dark place, after some time only the clean seeds are left behind.

Fig. The Resilient Tapered Thresher, Hand Model Made by Alf. Hannaford and Co. Ltd., Woodville, S. Australia, and used for Dry Zone Acacias.

Fig. CSIRO 15-cm Flailing Thresher, (A) Feeding Material into Thresher (B) General View Showing Material Threshed and Ready for Cleaning (C) View of Essential Parts.

Fig. Cement Mixer used for Dewinging.

Fig. *Liriodendron Tulipifera* Before and After Dewinging. Upgrading is Easier After Dewining.

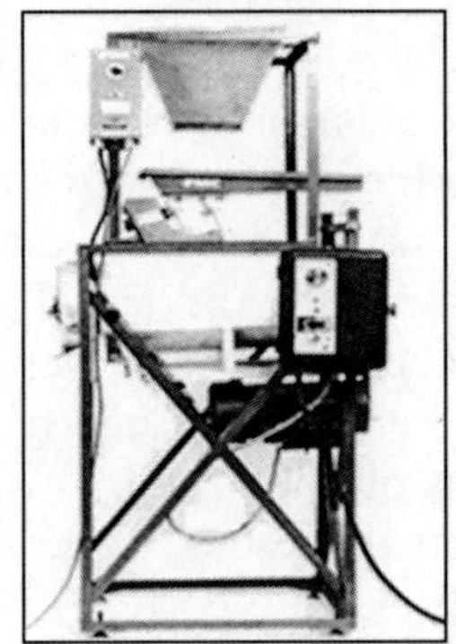

Fig. Missoula Dewinger for Small Seedlots.

Goor and Barney recommend that Cedrus seed be stored in the cones, since extracted seeds quickly lose their viability. When removed from storage, the cones must be soaked in water to facilitate extraction. After treatment the cones can be easily broken apart by hand and the seeds extracted for immediate sowing.

Serotinous cones of species such as Pinus brutia, P. halepensis, P. contorta and P. radiata may need special treatment to induce opening. Dipping them in boiling water for 10 to 120 seconds (up to 10 minutes for some especially refractory seed lots), followed by very high temperatures in the drying kiln (75° – 80° C), has been effective in some cases. High temperature is needed to melt the resin which forms a strong bonding adhesive between overlapping cone scales.

Immature green cones may also need special treatment. It was found that green cones of Pinus merkusii from Zambales (Philippines), after soaking for 48 hours, followed by 80 hours' drying at a starting temperature of 30° C and a final temperature of 50° C, released only 7% of the contained seeds; if the cycle of alternate soaking and drying was repeated 5 or 6 times, 79% of seeds were released. The total period of 4 – 5 weeks would be uneconomic in operational forestry. The recommended alternative is to collect only ripe brown cones which released 91% of their seeds in a single cycle.

OPERATIONS AFTER EXTRACTION

When seeds have been extracted from their fruits, several operations are needed before they are fit to go into storage. Sound seeds must be separated from empty and non-viable seeds and from inert fragments of fruits; winged seeds of some but not all species need to be dewinged; if seeds are to be stored, their moisture content must be tested and, if necessary, raised or lowered to the percentage most suitable for storage. If uniformity in growth of nursery stock is considered desirable, seeds may also be graded by size.

Inert material takes up space both in storage and transport and may cause uneven stocking in the nursery seed beds. It also carries a greater risk of introducing pests or diseases than do the seeds themselves; for example spores of needle cast are carried on needle fragments rather than seeds. Cleaning to a high standard of purity is easy in some species, but more difficult in others. Cleaning of seeds to a purity higher than a given percentage is undesirable in some species; beyond that, an increasing amount of good seed is separated out with the impurities. Besides, the added effort of extra cleaning is time-consuming and expensive. Morandini recommended that seeds ofLarix should not be cleaned to a purity higher than 65%, since it has been shown that further cleaning causes a drastic loss of good seed.

This is because the great thickness of the seedcoat in comparison with total seed size makes empty seeds nearly as heavy as full seeds. In many species of eucalypt, especially in the subgenera Monocalyptus and Idiogenes, it is difficult to separate the fertile seeds from the chaff which is released from the capsules at the same time. Viable seeds are frequently very similar in size, shape and colour to the chaff particles and the proportion by weight of chaff to viable seeds is usually in the range of from 5: 1 to 30: 1. Commercial seed lots of eucalypts are therefore cleaned of leaves, twigs and other large fragments but the remaining "seed" consists of a mixture of seed and chaff; this practice is accepted by both buyer and seller. Provided that seed lots are accompanied by test results giving the number of viable seeds per unit weight of seed plus chaff, the seed user will not be too concerned that they contain a certain amount of impurities. Cleaning seeds for operational use should therefore be applied with discretion. It may be necessary when special techniques are used *e.g.* if the seeds are to be pelletted or if precision sowing is to be employed. For research purposes also, when basic knowledge on germination or other seed characteristics is sought, repeated cleaning to a very high standard is necessary.

DEWINGING

Winged seeds or winged fruits are a feature of many forest trees and almost all conifer seeds have a wing which may vary from long and hard to very short and soft. In order to make seed processing and nursery sowing easier, the wing is usually removed whenever it is larger than the seed (or fruit).

Wings are small or impractical to remove from seeds in several coniferous genera *e.g.* Thuja, Chamaecyparis, Cupressus; in a few genera they cannot be removed without impairing seed viability *e.g.* Libocedrus. Many winged hardwood fruits *e.g.* Casuarina, Betula,Ulmus are stored and sown intact, but the larger wings of *e.g.* Swietenia fruits may be broken off.

For small quantities of seed dewinging may be done by hand, either by rubbing seeds between the hands or against a screen or roughened surface, or by handrubbing in a cloth bag or by rolling them between two cloth sheets or in a cloth bag between a rubber surface below and a roller above. For large quantities mechanical dewinging is in common use.

Dewinging machines range from those which are hand operated to large semiautomatic equipment which gives a continuous output. Corn mixers and cement mixers are frequently used. Mechanical dewinging, if carelessly done, may cause damage to seeds by crushing, cracking or abrasion.

Most mechanical dewingers are rotating devices in which the seed is pressed by brushes or pads against the walls of a cylinder, or rotating knobs or pads force the seeds to pass through narrow apertures while retaining the wings. If the clearance between the knobs or brushes is too small the seed may be damaged. A dewinger specially designed for small seed lots of 5 kg or less and for easy cleaning between seed lots has been described by Lowman and Casavan. It consists essentially of a rubber-lined cylinder with a rotating central shaft to which are attached pure gum rubber flaps. There is a variable inclination setting so that seeds and wings will pass out of the dewinger by gravity.

Nartov *et al.* describe a combined dewinging and seed cleaning machine in use in the USSR. It is transportable and weighs 50 – 70 kg. A seed hopper and feed auger pass seeds into the dewinging unit which has brush-type beater vanes. A fan then removes wings and lighter material, while the heavier material falls onto a series of inclined screens with various mesh sizes. The clean seeds fall into different containers, graded according to seed size.

Mechanical injury can be avoided in some cases by moist dewinging. Wang describes a safe method of dewinging coniferous seeds in Canada. The seeds are moistened with water and left to soak for 20–30 minutes before they are stirred with a soft brush or sponge in a rotating cement mixer to remove the wings. A similar principle is used by Isaacs (1972) in dewinging pine seeds. A large tank with slow-moving pipes gently agitates the seeds, which have been moistened at a rate of about 2 litres of water to 45 kg of seed. The wing absorbs moisture and is shed from the seed. In Honduras moist dewinging of Pinus caribaea and P. oocarpa is done through the use of a small cement mixer or rotating drum. The capacity of the drum should be at least double the amount of seed to be dewinged and the speed of rotation about 1 revolution every 2–3 seconds (Robbins 1983a,b). The winged seeds are tumbled dry in the drum for 15 minutes; water is then sprinkled slowly over the seeds as evenly as possible while the drum continues to rotate, at a rate of about 1 litre to every 50 litres of seed.

Tumbling is continued for 45 minutes after adding the water, and then the mixture is emptied out on to a gauze-bottomed tray, after which the seeds are separated from the wings. Moist dewinging is also in common use in Sweden. In moist dewinging the seeds absorb water and have subsequently to be dried to an acceptable moisture content.

OTHER CLEANING METHODS

A number of other methods of seed cleaning have been used experimentally, but are not yet in widespread operational use. They include electronic and electrostatic separators, magnetic separators, electronic colour separators and shaking tables, which separate seeds by the angle at which they rebound when thrown against fixed walls.

Seeds of Ochroma can be successfully cleared of their floss by placing the uncleaned mass on a wire sieve of 0.3 mm mesh and setting fire to the floss. The fire flashes across and the seeds drop through the mesh. Good results are obtained when the seeds are allowed to drop through the mesh into a pan of water. The inflammable oil in the floss burns with intense heat and experience in Honduras has shown that the floss must be spread out thinly to avoid damaging the seed. This method has also been tried to clear Populusseeds of floss but may damage more than 50% of the seeds in this genus.

Seeds of Prosopis are often embedded in a gummy matrix within the pods. *One way of obtaining clean seeds is:*

- To remove one side of the pod mechanically with a knife
- To soak the pod contents in a 0.1 normal solution of hydrochloric acid for 24 hours
- To wash in water for one hour, then dry in the direct sunlight
- To beat or pound the dried mass to separate the clean seeds from the gummy coating. This method has been successful in India, yielding clean seeds with a germination rate of 65% in 12 days.

4

Characteristics of a Seed Vigor Test

A vigor test should possess certain essential characteristics that can make it useful to the seed producer and consumer. characteristics

Inexpensive

Due to limited budgets for seed testing, it is important that a vigor test be reasonably priced and require a minimum investment in labor, equipment, and supplies.

Rapid

Every seed laboratory has periods of peak activity, thus it is important that the vigor test be conducted rapidly to minimize analyst time and germinator space. Furthermore, seed producers desire a quick turnaround time for samples submitted for vigor tests since such quick information on seed quality can provide them with a competitive marketing advantage.

Uncomplicated

Where possible, vigor test procedures should be simple so that they can be performed in seed laboratories without requiring additional staff with special backgrounds and training.

Objective

For a vigor test to be easily standardized, a quantitative or numerical index of quality that avoids subjective interpretations by analysts should be utilized.

Reproducible

The success of any test depends on its reproducibility. If these results cannot be repeated because of intricate procedures or subjectivity of interpretation, then comparison of results among laboratories becomes meaningless.

Correlated with Field Performance

Most definitions of seed vigor emphasize the relationship between seed vigor and field performance, and many studies have demonstrated this association exists. Consequently, the ultimate value of any vigor test may be its ability to predict field performance.

TYPES OF SEED VIGOR TESTS

The standard germination test is conducted under optimum conditions for seed germination. Consequently, when field conditions at planting are near optimum, the results usually correlate well with field emergence. However, under suboptimal field conditions, standard germination results usually overestimate field emergence. Therefore, additional tests are needed to better predict seedling emergence under a wide range of field conditions. Many vigor tests have been suggested; however, only a few have attained acceptance by seed analysts and seed testing organizations. This discussion will focus on the two most popular vigor tests: cold test and accelerated aging test.

COLD TEST

The cold test is one of the oldest methods of stressing seeds and is most often employed for evaluating seed vigor in corn and soybeans. Seeds are placed in soil or paper towels lined with soil and exposed to cold for a specified period, during which stress from imbibition, temperature, and microorganisms occurs. Following the cold treatment, the seeds are placed under favorable growth conditions and allowed to germinate.

Principles

The cold test exposes seeds to 10 C for 7 days in non-sterile field soil at approximately 60-70% of water holding capacity prior to a 4-7 day grow out period in ideal conditions (25 C). The moisture and temperature conditions provided in the cold test simulate the adverse conditions that seeds might encounter in an early spring planting. The ability of seeds to germinate and emerge in cold wet soil is affected by genotype, seed quality (both physical and physiological), seed or soil-borne pathogens and seed treatment. The cold test attempts to measure the combined effects of all of these factors. It often represents the lowest emergence that would be expected from a seed lot when planted under reasonably satisfactory field conditions, while the germination test represents the highest emergence potential that could be expected. When germination obtained in the cold test is very close to that obtained in the standard germination test, the seed lot would be expected to emerge well over a wide range of soil moisture and temperature conditions. There are two common approaches to conducting a cold test: rolled towel method and tray method.

Rolled Towel Method

Equipment and Supplies

Germination Paper Towels

Paper towels of the same weight and thickness as used for standard germination, 30 × 60 cm in size.

Plastic Bucket or Container

To hold rolled towels in an upright position for placement in germination chamber.

Plastic Bags (2 ml thickness)

To cover upright towels after placing in container.

Two Germinators

Two chambers capable of maintaining 10 C and 25 C (+/- 0.5 C).

Heat Exchange Plate

To maintain temperature to towels at 10 C.

Tray Method

Equipment and Supplies

Food Service Tray

Fiberglass, plastic or metal tray (45 x 66 x 2.75 cm).

Crepe-Cellulose Paper

Sheets of creped cellulose (40 x 61 x 0.5 cm) to fit in base or tray which is available from many paper supply houses.

Graduated Cylinder (500ml) or Automatic Watering Device

Used for adding specific amount of water to crepe-cellulose and tray.

Food Cart

Closed food service cart with plexiglass back to hold trays during testing.

Two Walk-in Temperature Controlled Chambers

Two chambers which allow entry of food service cart and are capable of maintaining 10 C and 25 C (+/-0.5 C).

Alternative Equipment

If laboratory does not have walk-in temperature chambers for food cart use, two germinators as described for rolled towels could be used and the trays moved manually from 10 C to 25 C.

Both Methods

Field soil: The soil component is a very important aspect of the test and the selection of an appropriate soil source is critical for the reproducibility of the test. Since the soil is expected to supply disease-causing organisms, it should be from a field site which has supported some vegetation, preferably the crop to be tested. Highly organic soils may exhibit lower germination due to excessive pathogenic activity which may cause erratic test results.

If possible, adequate soil for the entire testing season should be obtained from the same source. It should be handled in a consistent manner throughout the testing season and stored under uniform conditions to maintain soil moisture. Prior to use, the soil should be screened through a 5 mm sieve to remove large debris. It may be mixed with sand, vermiculite, peat or applied directly. Since the role of the soil in these tests is to provide an innoculum source, the quantity of the soil needed for each test is relatively small (i.e., one part soil/one or two parts sand).

Seed Samples

If cold tests are to be a useful guide in planting decisions, the seeds should be tested in the same condition as planted. They should not be allowed to dry out prior to testing, and they should have recieved their final fungicide treatment. Seeds at unusually low moisture levels (<8% moisture, fwb) may exhibit poor cold test performance due to imbibitional chilling injury; thus, seed lot moisture should be adjusted (if necessary) to a range of 8-13%.

ROLLED TOWEL

Procedures

This widely used procedure was first described in the 1950s. The towel procedure has gained widespread acceptance for its reliability, minimal space requirements, low volume of soil/sand needed, and generally good correlation with other cold test procedures and with field emergence. The common features of most of the current versions of this test recommend that the towel substrate includes soil or a soil/sand or peat mixture as an inoculum source. Following a low temperature (10 °C) stress period for 7 days, the rolled towels are transferred to an optimum temperature (25 °C) for germination for 5 additional days.

Methods

1. At least 16 hours prior to planting, the germination paper towels should be soaked until each towel has absorbed approximately 35 ml of water. The saturated towels should be equilibrated at 10 C overnight to ensure that the entire towel mass is at 10 C.
2. A minimum of four subsamples of 50 seeds each drawn from the pure seed fraction of each seed lot are tested. The seeds of each subsample are placed on a double layer of cold saturated towels in two 25-seed rows, 6 and 12 cm from the upper edge of the top towel. The planted seeds are lightly covered with soil/sand substrate (one part soil/one part sand) ensuring that all seeds are in direct contact with the soil. A top towel is placed over the two lower towels containing seeds and soil and the three towels are loosely rolled. Care should be taken during and after planting to ensure that the towels do not warm above 10 C. This can be done by placing the chilled towels in an insulated cooler over ice or by using a heat exchange plate.
3. When ready for transfer to the cold (10 C) chamber, the rolled towels are placed upright in a plastic bucket or container. The containers have wire dividers to keep the towels separated and are covered with a plastic bag to prevent loss of moisture.

4. The covered containers are placed in a cold room at 10 C (+/- 0.5 C) in darkness for 7 days.
5. After the cold treatment, the containers are transferred to a germination chamber at 25 C for 5 days in darkness.
6. Seedlings are evaluated using the same criteria as for the standard germination test

Tray

Common features of this test include a food service tray, a base of creped cellulose paper which acts as the water reservoir for the seed and a covering of a mixture of soil/sand. Test conditions are similar to the rolled towel test in that the first 7 days are at 10 C, but the growth period is shorter, 4 days at 25 C. Although the test requires considerably more space than the rolled towel procedure, it has several attractive features including: mechanisation and standardization of substrate watering, vacuum-head or trip-board planting, uniform distribution of seed onto the substrate allowing ease of interpretation, natural emergence of seedlings through a soil/sand substrate allowing faster evaluation and greater efficiency (less time per sample). The results of the Tray Method are correlated with field emergence and with other widely used cold test procedures. The primary limitations are the initial expense of the equipment and the relatively large amount of soil and floor space required to conduct the test, which may reduce the number of samples tested.

Proceduress

1. Prior to planting, the water, trays and carts and sand/soil should be chilled to 10 C.
2. One sheet of crepe-cellulose is placed into a standard food service tray and soaked with 1,100 ml of chilled water. The addition of water is most conveniently done with an automatic watering system which ensures uniform water distribution across the tray. Water can also be distributed in the tray with a graduated cylinder or other measuring device if desired.
3. Each tray accommodates 4 x 100 seed subsamples. Thus, two seed lots of 2 x 100 seeds each are planted in each tray. The 100 seeds in each subsample may be planted with a vacuum head or a trip planting board directly onto the saturated crepe-cellulose in each one-quarter section of the tray. The seeds are pressed uniformly into the crepe-cellulose using the bottom of an unused tray so that seed contact is constant and uniform distribution is maintained during the addition of the soil/sand substrate.
4. An excess of the chilled soil/sand mixture (1 part soil/2 parts sand) is spread over the seeds and crepe-cellulose. A board, notched to provide a level surface 0.4 cm below the lip of the tray, is used to level the soil/sand mixture and remove any excess. After the soil/

sand absorbs water from the wet crepe-cellulose, the resulting seedbed should approach 70% of the water holding capacity.

5. After planting, the test trays are placed at 7.5 cm intervals in a closed food service cart with the plexiglass back. The cart is moved into a 10 °C (+/- 0.5 °C) chamber in darkness for 7 days. The carts are then moved to a 25 °C chamber with an alternating light source (8-12 h light per day) facing the plexiglass back side of the cart for 4 days.

Interpretation of Results

The cold test does not provide an absolute vigor score, but records germination after 7 days stress under cold temperatures and high soil moisture in the presence of soil microbial activity. Since germination can be influenced by both the soil source and the cold test method used, the interpretation of results should be facilitated by the use of a control seed sample of moderate cold test performance with each seed lot or group of seed lots being tested. The relationship between the germination of the control sample and that of the samples being tested is probably more important than the actual germination of the test samples.

Within a laboratory and/or a seed company, the level of cold test germination can be related to field emergence and seed quality decisions can be made regarding potential seed lot performance. Until cold test methods are standardized, the results cannot be interpreted at a given vigor level among laboratories. The cold test is therefore limited primarily as an "in-house" vigor test within a laboratory and/or seed company.

Scope

The cold test has been used extensively for many years for testing corn and sweet corn seeds. The relationship between cold test results and field emergence has ranged from excellent to variable depending on the method used and the level of stress in the field.

When seeds are planted under stressful field conditions, however, there have been many comparisons which show higher correlations between cold test germination and field emergence than standard germination and field emergence. The general principles of the cold test have also been applied successfully to soybean and sorghum seed and could probably be applied to other kinds of seed.

The greatest difficulty with the cold test is the lack of uniformity in field soil. Soils differ in moisture, pH, particle composition, and pathogen levels, all of which contribute to divergent results. Use of vermiculite, a more uniform medium, has been suggested as a possible solution to the variability of soil conditions. However, it is widely believed that a cold test requires field soil to be successful. Regardless of these difficulties, cold test vigor rankings of seed lots tend to remain consistent within laboratories, which support this test as a most useful in-house vigor test.

Accelerated Aging (AA) Test

This test incorporates many of the important traits desired in a vigor test. Initially proposed as a method to evaluate seed storability, the accelerated aging test subjects unimbibed seeds to conditions of high temperature (41 C) and relative humidity (around 100%) for short periods (3 to 4 days). The seeds are then removed from the stress conditions and placed under optimum germination conditions.

Principles

The AA test exposes seeds for short periods to the two environmental variables that cause rapid seed deterioration; high temperature and high humidity. High vigor seed lots will withstand these extreme stress conditions and deteriorate at a slower rate than low vigor seed lots. Although aging temperature and seed moisture have the greatest influence on test results, several other factors which modify the results must be controlled during the test.

When conducting the AA test, seeds are weighed and placed on a screen tray, which is inserted into an inner chamber (plastic box) containing 40-50 ml of water. The inner chamber is placed into an AA (outer) chamber and the seeds aged at a specified high temperature for specified time, dependent upon the species. During the aging period, the seeds absorb moisture from the humid environment within the inner chamber and are stressed by high temperatures as seed moisture increases to a uniform level. It is important that seeds on the screen tray be kept in approximately a single layer at a uniform distance above the water surface, as the layering of seed will influence water uptake and result in variable seed moisture at the end of the test. Seed size will also influence the final seed moisture and germination following aging.

Larger soybean seeds had a lower final seed moisture and lower germination than smaller seeds when the same number were placed in the aging chamber. The final seed moisture can be controlled, however, by basing the sample size on a constant seed weight rather than seed number. Recommended and suggested testing variables have now been identified for a wide range of species A uniform temperature in the outer chamber must be maintained and the temperature controls should permit a variation of no more than +/- 0.3 C, since minor changes (0.5 C) in the aging temperature will affect germination results. High relative humidity in the outer chamber is also necessary to prevent water evaporation from the inner chambers.

A water-jacketed aging chamber is recommended because it provides precise temperature control, no condensation and a uniform aging environment. The door of the outer chamber must remain closed for the entire duration of the test to maintain uniform and constant temperatures. Opening the door for one minute lowered the temperature of the inner chamber 2-3 C and it took 30 minutes to recover to the desired temperature. The most uniform temperatures of the inner chambers are maintained by allowing some air space (2.5 cm) between inner chambers placed on the same shelf in the outer chamber.

EQUIPMENT AND SUPPLIES

Balance

Analytical balance capable of weighing to 1.0 mg.

Inner aging chamber

A plastic box (11.0 x 11.0 x 3.5 cm) with a lid into which is placed a plastic or wire tray with a 10.0 x 10.0 x 0.3 cm wire mesh screen (mesh 14 x 18). These trays can be purchased commercially.

Seed moisture tins

Temperatures ranging from 103-130 C as described for each species in the International Rules for Seed Testing

Bottle-top Dispensette (Brinkman)

Range from 0-100 ml, for dispensing 40 ml water from a standard screw-neck bottle or a 50 ml graduated cylinder.

Water

Deionized or distilled

Outer aging chamber

Water-jacketed incubator capable of maintaining a constant temperature range from 40-45 C +/- 0.3 C.

Alternative outer aging chambers

This may include incubators that have a heating element immersed in water at the base of the chamber. These chambers frequently do not have precise temperature control and require the use of auxiliary control to maintain uniform temperature (YSI model 71, Yellow Springs Instrument Inc., Yellow Springs, OH, USA). When using these chambers, water may collect inside the top of the outer chamber due to condensation and drip on the lids of the inner chamber (boxes).

If water accumulates on the lid of the inner chambers, condensation can occur inside the inner chamber (under the lid) and water will drop onto the seed that will raise seed moisture levels during the aging period. This can reduce germination and cause excess mold growth. Precautions must be taken to shield the inner chambers to prevent water droplets from accumulating on the lids during the aging period. Do not use dry incubators or ovens as outer chambers.

Germination test facilities

The brand name, model numbers and suppliers for outer and inner aging chambers and other equipment needed for the AA test can be acquired from the chairperson of the AOSA vigor test committee

Procedure

- The inner chamber AA boxes and screen trays should be thoroughly washed in a 15% sodium hypochlorite (clorox) solution and dried after each use to prevent fungal contamination.
- Place 40 ml (+/- 1 ml) of distilled or deionized water in each inner aging chamber and insert a dry screen tray, being certain not to splash water onto the screen surface.
- Determine the inital seed moisture of the sample to be tested using the method appropriate for the species and if >14.0% (fresh weight basis), dry the seed to 10 - 14% before testing.
- A minimum of 200 seeds, determined on a weight basis are placed on the surface of the screen tray (approximately one layer deep). More than one inner chamber should be used to obtain the quantity of seeds needed for larger seeded species). Preferably, seeds to be aged should be untreated. However, if seeds of the crop species are primarily marketed with fungicide treatment, treated seeds may be used.
- The lid is secured (not sealed) on each inner chamber. The inner chambers should be placed on a shelf, transported and placed in the outer aging chamber at the same time. Allow an air space of approximately 2.5 cm between inner chambers on each shelf in the outer chamber to assure temperature uniformity.
- Precisely monitor the temperature of the outer aging chamber and maintain temperature at +/- 0.3 C of desired temperature during the aging period. Record the time when inner chambers are placed in the outer chamber.
- The outer chambers should not be opened during the aging period
- Seed in the inner chambers should be removed from the outer chamber at the exact hour (+/- 15 min) specified and planted for standard germination within one hour after removal.
- The conditions for the standard germination test
- Include a control sample with each AA test. At the conclusion of the aging period, but prior to planting for standard germination, remove a small sample of seed (10-20 seeds) from the inner chamber of the control sample and weigh immediately to test for seed moisture (fresh weight basis) using the oven method. If seed moistures are lower or higher tested.
- When AA tests are initiated for many seed lots on the same day,

samples should be grouped at approximately one hour intervals between two outer chambers to allow adequate time for samples to be planted immediately after the aging period.

Interpretation of AA Results

The AA test does not provide an absolute vigor or field emergence score, but simply records germination (percentage normal seedlings) after a period of stress under conditions of high temperature and seed moisture. When the AA results are compared to the standard germination results of the same seed lot prior to aging, the AA germination will be either similar to standard germination (high vigor seed) or less than standard germination (medium to low vigor seed). One use of these results can be to rank seed lots by vigor and decisions can be made regarding the storability or planting of each seed lot.

Relationship with Field Emergence

Early storage studies suggested that AA might have utility as a vigor test to predict field performance. Additional studies have shown that this vigor test functions well in forecasting field emergence and stand establishment in a wide range of crop species. In general, when seeds are planted under stressful field conditions, AA germination provides higher correlations with field emergence than does standard germination.

Predicting storage potential

The primary emphasis of early investigations of the AA test was predicting the relative storability of seed of several crop species. It was shown that the changes in germination of high and low quality seed lots after a period of time were consistent with those that occurred in warehouse storage. Subsequent studies have verified that accuracy of this test in predicting the life span of a number of different species under a range of storage conditions.

Limitation of AA

Accelerated aging shows great potential as a vigor test for a wide range of species. Although variability in referee results was shown initially, excellent progress has been made in the refinement of techniques and the evaluation of time, temperature and seed moisture interactions. AOSA and ISTA referee tests with soybean seed report greater uniformity among laboratories and confirm the relationship between AA germination and field emergence. Sufficient evidence has now accumulated to show that repeatable results can be achieved using this test.

Thus, the AOSA and ISTA now consider this test standardized and have recommended it as a vigor test for soybean.. Even slight modifications in temperature control, sample size or aging time will cause variation in final seed moisture and/or germination which will limit the acceptance of the vigor test. The accelerated aging test is rapid, inexpensive, simple and useful for all species; it can be used for individual seed evaluation and requires no additional training for correct evaluation.

INTERPRETATION OF SEED VIGOR TEST RESULTS

It is extremely important to recognize that the results of a seed vigor test do not predict percentage field emergence. This is because environmental conditions differ from field to field, day to day and year to year, and a particular seed lot will reach a different emergence percentage under each set of environmental conditions.

Seed vigor tests supply only relative values. This means that tests must be conducted on a number of seed lots simultaneously with the inclusion of a control seed lot with known seed vigor to serve as a reference point. This allows the arrangement of seed lots in ranking order, from high to low vigor. Because plant stand will vary from field to field, and because vigor tests, in some cases, vary among laboratories, it is very difficult to establish cut-off points between acceptable and unacceptable levels of seed vigor across many crop species.

Many seed companies in the central United States conduct a series of vigor tests on each soybean seed lot and combine this information into a seed vigor index. They establish an acceptable vigor level for a specific production season and seed lots are evaluated (in-house) prior to conditioning, treatment, and marketing. In a few cases, those seed lots that exceed a specified vigor level after conditioning are advertised and promoted in general terms as vigor-proven, vigor-rated, or high vigor seed. If the seed is sold with strong emphasis on vigor, a buyer should ask the dealer for specific information on how many vigor tests were conducted and the criteria used to assess vigor potential.

Does high-vigor seed mean higher yields? This question is asked by many farmers following implications concerning yield by some seed producers. Unfortunately, there have been fewer comparisons of seed vigor to final yield than to field emergence. As with other investigations relating to seed vigor, the yield comparisons are variable depending on the crop planted, the stand achieved, and the original vigor level of the seed. Many factors that may not relate to the vigor of the seed can influence yield during the growing season. Also, certain crops and genotypes have the ability to compensate for minor stand differences following emergence, and this may result in little difference in final yield. As would be expected, if adequate field stands resulted from high vigor seeds, higher yields should occur. However, less evidence supports improved yields using high vigor seeds than supports equal stands achieved from planting medium or low vigor seeds. To date, insufficient evidence is available to show a strong relationship between yield and vigor in the absence of stand differences. Future research on plant growth and development or the physiological or biochemical basis for seed vigor may improve our understanding of this relationship.

SEED HABIT TEST

The seed habit is the most complex and evolutionary successful method of sexual reproduction found in vascular plants. Today, seed plants, gymnosperms and angiosperms are by far the most diverse lineage within the vascular plants. Most of this diversity is accounted for by the angiosperms and Charles Darwin described the rapid rise and early diversification within the angiosperms during the Cretaceous time as 'an abdominable mystery'. Imagine such a plant, producing only one megaspore per megasporangium. The megaspore is not released until the female gametophyte has fully developed within it, complete with archegonium and egg.

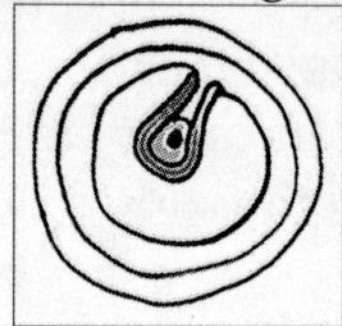

Imagine that the whole megasporangium (with its single megaspore) is released. Imagine too that the megasporophyll and ligule (A, B) extend and enwrap the megasporangium, protecting it. This composite structure is an ovule.

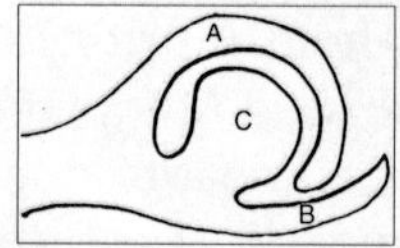

C = megasporangium

Imagine that this protected megasporangium (with its single megaspore and internal gametophyte) is retained on the parent sporophyte until the egg in its archegonium is fertilized and forms an embryo. Only when this has happened, and food reserves are laid down with the embryo, and the outer protective integuments harden, is this structure released. This is what we call a seed.

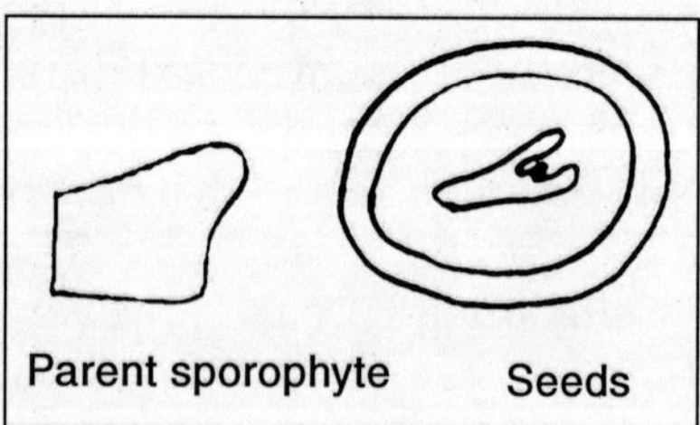

EXISTENCE

The seed plants have an adaptive advantage, occur in a wide variety of habitats and dominate today's flora. This evolutionary success is, at least in part, due to the seed, which is one of the most dramatic innovations during land plant evolution. Morphological evidence for the monophyly of seed plants includes the seed habit itself, but also include vegetative traits like the production of wood by cambium, a secondary meristem. The origin and evolution of the seed habit is a fascinating story by that started in late Devonian times about 385 MYBP. To understand the seed, it helps to think about how it evolved and what it essentially is in terms of origin and function.

The earliest seed plants, "progymnosperms", emerged in the late Devonian. Progymnoperm fossils show vegetative morphologies to seed plants, but not all progymnosperms had seeds or seed-like structures (ovules or pre-ovules). Archaeopteris spp. was the first modern tree, but it sproduced spores rather than seeds. However, it exhibited an advanced system of spore production called heterospory. Heterosporous plants produce two sets of specialized spores: megaspores (haploid female-like megaspores) and microspores (haploid male-like microspores). Heterospory, which probably has been evolved independently in several lineages, is widely believed to be a precursor to seed reproduction.

The progymnosperms are regarded as the ancestors of the seed plants. Fossils of seed-bearing seed ferns (Lyginopteridopsida) exhibit a variety of seed and seed-like structures (see 'The earliest seeds'). 'The seed' might have evolved once or several times during evolution.

Three major evolutionary trends were important for the transition from the seed ferns to the gymnosperms, from the spore to the gymnosperm seed:

1. The evolution from homospory to heterospory and connected with this from megasporangia with many spores to megasporangia with just one functional megaspore.
2. The evolution of the integument, maternal tissue that protects the ovule; the integument forms the seed coat.
3. The evolution of pollen-receiving structures. This includes the transition to water-independence of the pollination/fertilization process (water is required for fern fertilization).

Today there are four major lineages of extant gymnosperm seed plants: Cycadopsida (cycads, "Palmfarne"), Ginkgopsida (ginkgos), Pinopsida/ Coniferopsida (conifers, "Nadelbäume"), and Gnetopsida (gnetophytes). Extinct gymnosperm groups include the Lyginopteridopsida (seed ferns, Pteridosperms, "Samenfarne", paraphyletic group including Devonian/ Carboniferous Lyginopterids and Carboniferous/Permian Medullosans), Bennettitales (cycadeoids), Gigantopteridales (gigantopterids), Pentoxylales (*e.g.* Pentoxylon), Caytoniales (*e.g.* Caytonia), Glossopteridales, Voltziales (*e.g.* Emporia), and Cordaitales.

The evolutionary connections between these gymnosperm groups are uncertain and especially the position of the Gnetales is a matter of controverse debate. Based on molecular (DNA sequences), morphological (including fossil seeds), and biogeochemical evidence (oleanone) the different gymnosperm groups are either monophyletic or paraphyletic, and their evolutionary relationships to the angiosperms are unclear. Anthophyte hypothesis: Virtually all morphology-based cladistic analyses propose a monophyletic group consisting of angiosperms + Bennettitales + Pentoxylales + Caytoniales + Glossopteridales + Gnetales (= anthophytes). This implies that the Gnetales are sisters of angiosperms and the gymnosperms are paraphyletic. This hypothesis is not supported by molecular data, which place the Gnetales as sister to all remaining seed plants, and in several molecular analyses even within the conifers as sister to Pinaceae. Based on many molecular analyses, the gymnosperms are a monophyletic sister group of angiosperms.

Among several other possibilities, the Caytoniales, Pentoxylales, Glossopteridales, Bennettitales and Gigantopteridales have been proposed as ancestors of the angiosperms. Gigantopterid hypothesis: The Bennettitales (cycadoeoids), which are morphologically highly similar to the Cycadopsida (cycads), are a group of gymnosperms that are highly similar to angiosperms. Oleanane, a triterpene important for insect defence is widespread among angiosperms, has been discovered in fossils of Cretaceous Bennettittales (cycadoeoids) and Permian Gigantopteridales but not in other gymnosperm groups. The gigantopterid hypothesis claims that the angiosperms derived from seed ferns via cycadoeoid-related gymnosperms (gigantopterids).

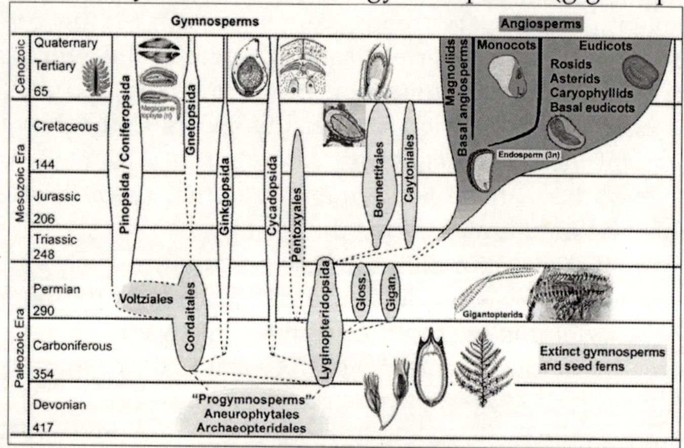

INSIDE OF SEED

A seed consists of an embryo, stored food and a seed coat. The seed replaces the spore of the seed-less fern plants as propagation, dispersal and deposit/outlast/storage unit.

Ferns and seed plants both exhibit a life cycle in which two heteromorph generations alternate:

1. The dominant diploid sporophyte, which is the fern or spermatophyte plant that you actually see as the 'large-sized' organism in nature.

2. The haploid gametophyte, which for the ferns is the small-sized (few mm to few cm) prothallium of ferns that you might see in nature

In the seed plants the haploid gametophytes became a 'hidden generation' that completely depend on the sporophyte, it was hidden to us until Hofmeister discovered the alternation of generations in seed plants in 1851. In seed plants the male gametophyte (microgametophyte) is hidden in the pollen grain ("Pollenkorn"), and the female gametophyte (megagam-etophyte) is hidden in the ovule ("Samenanlage"), after pollination and fertilization the ovule develops into the seed. Angiosperm and gymnosperm gametophytes and seeds are distinct. To understand the angiosperm seed, it is necessary to understand the gymnosperm seeds, and how it evolved from seed ferns and progymnosperms. The gymnosperms have 'naked seeds', *i.e.* their ovules and seeds (fertilized ovules) are exposed on the surface of sporophylls and analogous structures. The megagametophyte (female gametophyte, n) develops from the functional megaspore (n) within the nucellus (megasporangium, 2n). The megagametophyte of the gymnosperms is homologous to the megaprothallium (n) of the ferns and is sometimes called primary endosperm (n). The megagametophyte of seed plants is retained and nourished by the parent plant within the ovule. Ovule = megagametophyte + megasporangium (nucellus) + integument (seed coat).

The megagametophytes of the gymnosperms produce several archegonia (n) with egg cells (n). Fertilization by the sperm from the pollen grain (microgametophyte, male gametophye, n) often leads to the development of several embryos (2n) within a single ovule. Polyembryony of gymnosperm seeds is a known phenomenon. In most cases only one embryo survives and therefore relatively few fully developed gymnosperm seeds contain more than one embryo. In contrast to the gymnosperms ("Nacktsamer"), the angiosperms are "Bedecktsamer", *i.e.* their ovules and seeds are enclosed inside the ovary, which is the base of a modified leaf and is called carpel. Another very important difference to gymnosperms is the angiosperm double fertilization. This leads to an additional novel tissue with maternal protuberance, the triploid endosperm. In mature seeds of most angiosperm species, the embryo is enclosed by endosperm tissue. In addition, angiosperm seeds can be dispersed as fruits, *i.e.* the seeds can have in addition pericarp (fruit coat) around the testa (seed coat). The Triassic and Jurassic age was dominated by gymnosperms, although the first angiosperms evolved during at this time. The rapid rise and early diversification of the angiosperms occurred during the Cretaceous time and was called 'an abdominable mystery' by Charles Darwin.

THE EARLIEST SEEDS

PROGYMNOSPERMS

The earliest seed plants emerged in the Devonian, and have been called "progymnosperms". Progymnoperm fossils show vegetative morphologies to gymnosperms: They were shrubs or trees with laminate leaves, but reproduced by spores like ferns.

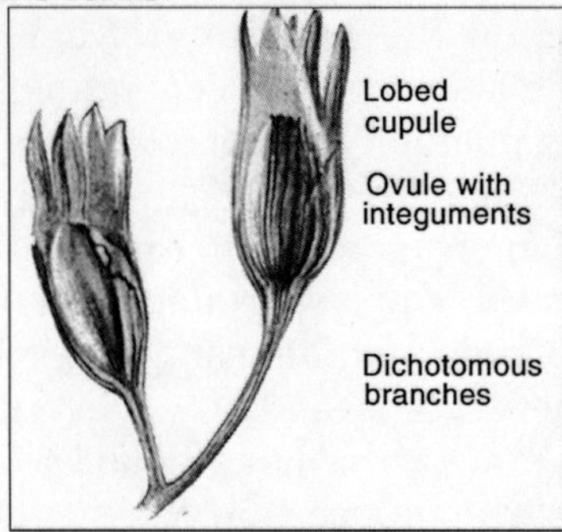

They had a bifacial vascular cambium that produced secondary xylem (wood with pitted tracheids) and secondary phloem. The progymnosperm Archaeopteris spp. was the first modern tree, but it produced spores rather than seeds. However, it exhibited an advanced system of spore production called heterospory.

Heterosporous plants produce two sets of specialized spores: megaspores (haploid female-like megaspores) and microspores (haploid male-like microspores). Heterospory, which has been evolved independently in several lineages, is widely believed to be a precursor to seed reproduction. Thus, both the production of wood and heterospory predate the evolution of the seed. The progymnosperm groups of Archaeopteridales and Aneurophytales are regarded as the ancestors of the seed ferns and seed plants.

DEVONIAN SEED FERNS

The oldest fossils of ovules ("Samenanlagen") or seeds (fertilized, mature ovules) are from the late Devonian (>365 MYBP, Runcaria even 385 MYBP). The earliest seed plants with seeds or seed-like structures are Devonian seed ferns (Lyginopteridposida, Pteridosperms). Several different types of preovules or preovule-like structures are known. Not all characteristics of the seed habit (heterospory, a single megaspore within a megasporangium (nucellus), megasporangium enclosed by an integument, pollen capture before seed dispersal) are evident in these preovules. Runcaria (Gerrienne *et al.* 2004), the oldest seed-bearing seed fern (Middle Devonian, 385 MYBP) had a small, radially symmetrical, integumented megasporangium surrounded by a cupule. The megasporangium bears an unopened distal extension protruding above the multilobed integument. This extension is assumed to be involved in wind pollination (anemophily).

Runcaria sheds new light on the sequence of character acquisition leading to the seed. In general, a seed is simply a mature ovule containing an embryo. An immature ovule consists of a diploid megasporangium (nucellus), containing a single functional megaspore that develops into a haploid megagametophyte. The megasporangium is surrounded by diploid covering layers, the integuments (which evolve into the seed coat). An integumentary opening at the apical end is important for pollination, wind pollination in seed ferns and modern gymnosperms; this opening evolved into the micropyle.

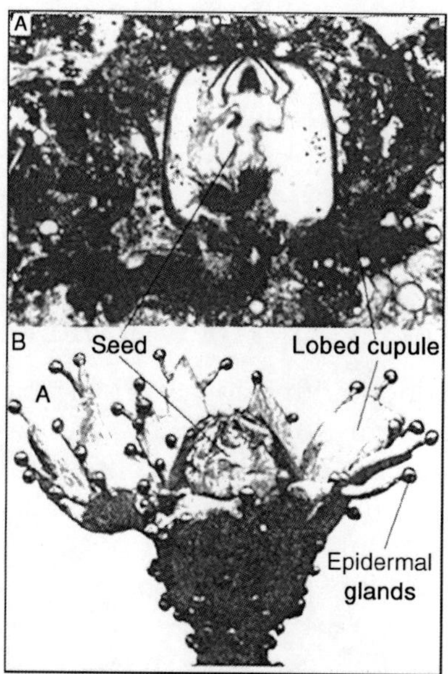

Fig. Fossil (A) and Rconstruction (B) of a Late Devonian Lyginopterid-type Seed of Lagenostoma Lomaxi.

The Late Devonian and Early Carniferous seeds ferns are sometimes collectively called Lyginopterids and include Elkinsia, Archaeosperma, Lagenostoma, Moresnetia, and Lyginopteris. These earliest seed-bearing seed plants produced their preovules or ovules on dichotomously branched, sterile structures called cupules.

Cupules are cup-like structures that partially enclose the ovule. In these early ovules the nucellus was surrounded by integumentary tissue consisting of more or less free lobes. The integumentary lobes curved inward at their tips, forming a ring around the apical end. The integuments of the ovules evolved through gradual fusion of the integumentary lobes. The integuments later evolved into the seed coat. An opening that was left at the apical end, evolved into the micropyle, it permitted pollen to enter and to fertilize the egg cell. The earliest seed plants employed "hydrasperman reproduction". Pollination occurred when wind-blown pollen was directed into a semi-closed pollen chamber. In several cases a specialized lagenostome was used for pollen capture. A lagenostome is a funnel-like structure of the nucellus that projects from the top of the megasporangium, it functions as a trumpet-like pollen-trapping device. From there the pollen was delivered to the pollen chamber. A central column (lagenostome column) was attached to the pollen chamber floor sealed the chamber to provide optimal conditions for pollen germination.

As the megagametophyte matures the opening to the pollen chamber is sealed and the floor is ruptured, allowing fertilization to proceed. The germinated pollen grain (microgametophyte) settles into an intimate proximity to the ovule and delivers the sperm/spermatozoid (has cilia, swims, requires water) into the archegonia with the egg cells.

Fossil records suggest that the sperm delivery required lysis of the megasporangium wall. The important issue was that pollination and/or fertilization became more water-independent during evolution, which facilitated the diversification of seed plants from Carboniferous through to the present day. So far, embryos have not been found in Devonian seed fern fossils.

MEDULLOSAN SEED FERNS OF THE CARNIFEROUS AND PERMIAN

The seed ferns (Lyginopteridposida, Pteridosperms) are a paraphyletic group of extinct gymnosperms. An interesting group, the "medullosan seed ferns" were abundant trees in Carboniferous floodplains end extend well into the Permian. This group includes Trigonocarpus, Pachytesta, Rhynchosperma, Medullosa and Stephanospermum.

Seed-bearing Seed Fern Emplecopteris Triangularis from the Permian of China.

Fossil seeds from medullosan seed ferns (see figures on the left and below) are several mm to several cm long. In some cases even embryo structures have been preserved. The ovules are usually radiospermic, with one end of the integument drawn out into a micropyle that probably helped guide pollen to the megagametophyte within. A pollination drop mechanism may also have aided pollen capture. A small pollen chamber appears just inside the micropyle. This structure is preserved in detail in a number of Devonian and Carboniferous seed fossils. A loss of the cupule and a loss of the lagenostome column was evident in the medullosan seed ferns.

The integument of Pachytesta and Stephanospermum is three-layered with an epidermis and outer fleshy layer (sarcotesta), covering a though fibrous (sclerotesta) and thin endotesta, which lays adjacent to the nuclleus (see figure on the left). In Stephanospermum konopeonus ovules (Drinnan *et al.* 1990) an apical funnel aided wind-blown pollen capture. Beyond the micropyle and tip of the sclerotestal beak, the sarcotesta flares open as a funnel that tapers to the micropylar opening at the tip of the sclerotestal beak. The rim of the funnel forms he broad apex of the ovule.

The endotesta of Stephanospermum ovules is thin and lines the inner surface of the sclerotesta. The nucellus and the megagametophyte are poorly perserved in the Stephanospermum ovule fossils. In these ovules, the nucellus may have consisted of only a few layers of thin-walled parenchyma surrounding the megaspore membrane. The nuclleus is attached to the integument only at the base of the ovule. The megaspore membrane is robust and consists of a distinctive network of granules and rods of sporopollenin covered by a homogeneous outer layer. Cells of the megagametophyte are not preserved from Stephanospermum ovule fossils.

STANDARDIZATION OF VIGOR TESTS

For any seed quality test to be useful, it must provide reproducible tests. To evaluate reproducibility of various vigor test results, both the AOSA and ISTA have conducted extensive referees. Generally, results of these referees have shown that seed testing laboratories can reproduce their own results on the same sample of seed within acceptable confidence limits. However, when the same tests are conducted by different laboratories, the amount of variability can be unacceptable.

Many possibilities exist to explain the lack of standardization among laboratories. For example, most vigor tests require a degree of subjectivity. The seedling vigor classification and tetrazolium vigor tests possibly involve the most subjective interpretations because seeds and seedlings or both are separated into categories based on characteristics that are difficult to describe precisely. Even the cold test and accelerated aging test require the classification of seedlings as normal or abnormal.

Variations in temperature, moisture, and other environmental conditions are more critical for tests in which rate of growth or rate of a biochemical process is measured than for those such as the standard germination test which measures the completion of a process. For example, a 1 C difference in temperature during the course of the standard laboratory germination test probably has little effect on the final percent germination, but a 1 C temperature difference may have a considerable effect on results in the seedling growth rate test or seed deterioration in the accelerated aging test, as would minor variations in the moisture content of germination substrata or the relative humidity of the air. Consequently, conditions and equipment that are suitable for the standard laboratory germination test may not be suitable for vigor tests.

The cold test would appear to be difficult to standardize because it exposes seeds to soil microorganisms to measure their ability to resist attack under cold, wet conditions. Thus, the standardization of the cold test procedure in its present state still requires standardization of the substrate (soil) microflora, a task that would be difficult, if not impossible. The use of sterile media inoculated with specific microorganisms has been suggested, an approach that is probably too simplistic, since it is difficult to culture microorganisms and maintain a constant level of pathogenicity. Also, pathogens behave differently in soil, where there is a complex population of microorganisms, from how they behave in the absence of such an interacting population. Despite the problems, the cold test for corn in the U.S. corn belt is as repeatable as the standard germination test. .

It may be important to determine whether such factors as dormancy affect vigor test results. Many species (e.g., legumes) have seeds that are impermeable or only slowly permeable to water. This can affect the percentage of seeds that germinate as well as the length of seedling growth in stress tests and can cause a biased index of vigor in a seedling growth rate test. It can also affect the leaching of electrolytes from seeds in a conductivity test.

Dormancy can also confound the interpretation of vigor test results. For example, dormancy can cause lettuce and cereal seeds to germinate poorly in tests conducted at high temperatures. If dormancy is suspected of affecting vigor results, the seed should be checked by a standard warm germination test or a tetrazolium test.

CONTROL SAMPLES IN VIGOR TESTING

An essential component of any laboratory testing program is the use of control or check samples. The primary purpose of control samples is to provide internal quality control to enable seed technologists to detect variations in test procedures. Seed vigor tests require precise time, temperature, and moisture that must be monitored and controlled during testing. Even small variations in these components can produce dramatic fluctuations in test results. The selection, maintenance, and use of control samples for each species tested is essential for laboratories engaged in vigor testing.

Control seed lot Selection

Seed lots of each crop species tested should be selected annually by screening several samples for both germination and vigor. The sample selected should be relatively free of physical injury and disease, high in germination and moderate in vigor. Seed lots with extremely high vigor will not allow detection of minor differences in methodology and, therefore, would not make good control samples. For example, an acceptable control sample for corn or soybean may have a germination exceeding 95 or 90%, respectively, and vigor level of 70 to 80% and 80 to 90% for the accelerated aging test for soybeans and the cold test for corn, respectively.

Seed Storage and Maintenance

Store enough seed of the control sample to provide an adequate supply for an entire year or testing season. Store the control samples in conditions that will preserve both germination and vigor. Prepackaging the control sample into sub-sumples needed each day may be advantageous, removing from storage the number of samples needed for a week of testing. Two alternatives should be considered for storage and use of control (in order of preference):

Divide the control seed lot into many subsamples of ~500 seeds. Each subsample should be packaged in moisture-proof foil packets (heat sealed) or stored in a moisture-proof container at low temperatures (0 to -20 C) prior to testing. Control subsamples must be allowed to equilibrate to room temperature (4 to 6 hours) prior to use.

If it is not possible to handle the small control samples described above, subdivide the control sample into larger subsamples. Place these samples into moisture-proof containers and store at low temperatures (0 to 10 C) prior to testing. Remove small subsamples from the larger samples as needed for testing, place in sealed containers and allow to equilibrate to room temperature before use.

Seed moisture of control samples

The initial seed moisture of the control seed lot must be an acceptable seed moisture for the seed being tested. The moisture of the seed lot must be measured periodically to be certain that the seed has not decreased or increased in moisture during storage. Adjust the moisture to the initial level if necesary, as extremely dry seed can cause imbibitional injury and inaccurate vigor readings during testing.

Use of control samples

Control samples should be included as routine samples in every vigor test run and should not be identified as control or check seed until the test is completed. At the conclusion of the test run, the seed technologist or laboratory manager should determine if the vigor test results of the check are within the set parameters. If the sample falls outside the parameters, the problem should be identified and corrected and the test repeated. At frequent intervals, the vigor test results of the control samples should be summarized to evaluate uniformity of the samples and testing procedures.

Clearly, vigor testing for all species and regions has not yet achieved the same level of standardization possessed by the standard germination test. Considering the important role that vigor tests have in plant breeding, seed production, quality control, and marketing programs, standardization is badly needed. Although future research and testing are required before vigor testing becomes a routine phase of seed testing, the promise of vigor testing in the future is bright.

USE OF SEED VIGOR TESTS

Many farmers know from experience that seed lots with equal germination levels may emerge from the soil quite differently, resulting in erratic field stands, replanting (on some occasions), or both. These same farmers recognize that this problem is usually more severe when adverse environmental conditions occur at or immediately after planting. Thus, if farmers know that they will be planting seed in adverse soil conditions or can predict an adverse field environment following planting, high vigor seeds should provide higher field emergence than low vigor seeds. If this prevents replanting, and the associated delayed maturity or yield reductions due to poor stands, additional money spent on high vigor seeds may be a worthwhile investment. It cannot be assumed, however, that high vigor seeds will produce excellent emergence and stands in any soil environment. It should improve the chances for satisfactory emergence, but it will not guarantee it. Conditions prevailing after planting may be so severe that even the most vigorous seeds cannot perform satisfactorily.

5

Pathogens of Seed Borne

PATHOGENS ASSOCIATED TO THE SEED

As an introduction we will remind some aspects of the morphology and anatomy of the seed, in order to understand in a better way the effect pathogens have in the seed and later, the plant originated from it.

THE SEED

The seed consists of three basic parts:

- Embryo,
- Storage tissues
- Seed coat.

The embryo has a reproductive function, being capable of initiating cellular divisions and growth. It is the most important part of the seed. It is an axis that originates growth in two directions, with the objective of originating a root and a shoot. Usually, the embryo is very small, compared to other parts of the seed. It is like a miniature plant. It is consists of a radicle, plumule, one or two cotyledons, and hypocotyl or epicotyl, depending on the type of plant.

Energy storage is localized in the cotyledons, in the endosperm or the perisperm. Cotyledons originate from the zygote and are part of the embryo. In many species, storage substances are localized in the cotyledons, and the embryo develops absorbing all the endosperm.

Episperm is the seed cover, consists of two layers, the testa and the endopleura. The outer layer is the testa; it can be stony, leathery, membranous or fleshy. Over the testa we can recognize: the hilum, scar or point of attachment of the seed to the funiculus, water penetrates easily through it; Micropyle, the point upon the seed at which was the orifice of the ovule through which the pollen tube enters; raphe, suture originated by the part of the funiculus that is fused along the side of the ovule. The endopleura is the inner layer; it is thin and generally whitish. Teguments, testa or protective covers delimit the seed. They are formed by one or more layers of cells originated from the ovule integuments and sometimes from the pericarp made from the walls of the ovary. Regarding the healthiness or pathology of the seed we will consider three aspects that allow us to understand processes associated to seed-borne diseases: infection of the seed, infection of the plant and steps that can be taken to reduce the damage caused by this relationship.

Seed Infection Mechanisms

The area of science that studies the relationship between pathogens and seeds is Seed Pathology. It does not only identify the pathogens, it also includes the role of the seed as source of inoculum, the survival of the pathogen and the actions taken to control the pathogens associated to it. It uses the knowledge of General Pathology, Microbiology and Seed Analysis. To obtain a perspective of seed borne diseases, seed borne microorganisms can be considered fewer than four classes.

The first consists of pathogens for which the seed is the main source of inoculum; when seed infection is controlled, the disease is controlled. An example would be lettuce mosaic virus. For these pathogens, the importance of seed-borne inoculum has long been recognized, and control practices have been developed. The second class consists of important pathogens in which the seed borne phase of the disease is of minor significance as a source of inoculum.

Examples are those in which the crop residues in the field were the major source of inoculum. The third and largest group of seed borne microorganisms consists of those that have never been shown to cause disease as a result of their presence on seeds. The fourth class is a group of microorganisms that can infect the seed either in the field or in storage and reduce yield and seed quality. Examples of field fungi are *Diplodia, Fusarium, Cladosporium,* etc.

The storage fungi *Aspergillus* and *Penicillium* can invade most types of seeds under high-moisture storage conditions. The process of seed infection is influenced by the conditions under which the crop grows. Between the facts that influence in the process of infection are: the host and its genotype, the pathogen and its genotype end environmental facts. An infected seed will not always an infected seed will be the cause of the infection of the infection in the plant which originates, so, it is to say that the infection caused from an infected seed is an exception, not something to happen generally. There are two situations, Systemic infection of the seed and Contamination or Infestation of the seed.

Systemic Infection of the Seed

The establishment of a pathogen in any part of the seed is refereed as seed infection. It can be systemic, by the vascular system or plasmodesmata or directly by natural or artificial wounds. The same pathogen can infect the seed using one or more of these mechanisms. For example: *Xanthomonas campestris* pv. *phaseoli* can infect seed through the vascular system, by natural openings (From the pod suture goes to the funiculus, then to the raphe and tegument, or it can also happen thought the micropyle).

Systemic Infection through Flowers, Fruits or Funiculus

Most of the systemic seed-borne bacteria and fungus reach and infect the embryo through the flower or from the peduncule of the fruit, via funiculus. Viruses go to the embryo from the systemically infected mother plant and the infected or contaminated pollen. They rarely reach the embryo during the formation of the seed or the embryo itself.

Examples of some infections that occur through the vascular system are: Some *forma specialis* of *Fusarium oxysporum* in pumpkin, pea and tomato; *Plasmopara halstedii* in sunflower; *Septoria glycines* in soy; *Verticilium dahliae* in spinach and sugar beet; Certain pathovares of *Xanthomonas campestris* in bean, cabbage, rice and sweet pepper; and *Pseudomonas syringae* pv. *Lachymans* in cucumber.

Penetration through the Stigma

During the infection, some pathogens follow the same path as the pollen grains do. Spores of some fungus reach the stigma and germinate, producing an hypha that reaches the ovary through the style, where they can stay as a dormant mycelium until seed germination.

For example: *Ustilago nuda* and *U. triticci* in barley and wheat, and *Alternaria alternata* in sweet pepper. Viruses can infect through infected pollen, the male gamete carries the virus, when joining the ovule it generates an infected embryo. If both, the male and female gamete are infected they can even produce an infected endosperm.

Penetration through the Wall of the Ovary

Some fungi, like *Ustilago nuda* and *U. triticci* penetrate through the wall of the ovary as a result of the germination of the Teliospores on the stigma or the wall of the ovary. The pro-mycelium goes through the wall and other tissues until it reaches the embryo. In some other cases, penetration occurs through breakages on the testa, establishing itself in the endopleura or the endosperm. In fleshy fruits, like cucumber, melon, eggplant, tomato, sweet pepper and others, contamination can occur directly through the funiculus or in the tegument, during the process of seed formation.

Examples of this are *Colletotrichum lagenarium* in watermelon and *Rhizoctonia solani*, when it invades fleshy fruits, it is capable of infecting from the placenta and penetrate to the developing ovule or seeds that are still in its formation process and have not lignified its cover.

Penetration through Wounds and Natural Openings

Natural openings like the hilum and the micropyle or wounds generated during the thresh are spots where pathogens like *Xanthomonas campestris* pv. *phaseolicola* in bean and *Pseudomonas syringae* pv. *lachrymans* in cucumber.

SEED CONTAMINATION OR INFESTATION

Contamination or infestation refers to the passive relationship of a pathogens and seeds.

The pathogen itself or parts of it can stick or can get mixed with the seeds during any of the processes during seed recollection: harvesting, extraction, thresh, selection and packing.

Pathogens that Stick to the Surface of the Seed

Pathogens that stick to seeds during harvest or postharvest do so by their spores (Clamidospores, Oospores, Teliospores, Uredospores), bacterial cells and in some cases, virions. Spores of the following fungi can be carried on seed coat surfaces: *Alternaria brassicae and A. brassicicola* in crucifers; *A. longipes* in tobacco; *A. radicina* in carrot; *Ascochyta pinodella* in pea; *Drechslera sorokiniana* and *D. oryzae* in rice; *D. avenae* in oats; *Fusarium oxysporum* f. sp. *callistephi* in China aster; Sclerotia of *Rhizoctonia solani* in eggplant, pepper, and tomato; *Tilletia caries, T. foetida, T. contraversa*, and *Urocystis agropyri* in wheat.

A number of bacteria, such as the following, contaminate seed surfaces: *Corynebacterium flaccumfacjens* pv. *flaccumfacjens* in bean, *Pseudomonas syringae* pv. *phaseolicola* in bean, *P. syringae* pv. *lachrymans* in cucumber, *P. syringae* pv. *tomato* in tomato, *C. rnichiganense* pv. *michiganense* in pepper, *Xanthomonas campestris* pv. *campestris* in cabbage Also some viruses as Tobacco mosaic virus, Tomato mosaic virus, Pepper Mosaic virus.

ACCOMPANYING CONTAMINATION

This type of contamination refers to physical mixing of the seed with pathogen's propagation organs like the sclerotium, nematode's galls, contaminated plant parts or soil particles containing pathogens.

Structures of the Pathogens

Some fungi produce resistance structures named sclerotiums that are made of compacted mycelium. Under certain conditions of temperature and humidity it can germinate, it is common that this occurs together with the seed, producing its infection, by direct penetration or by the spores generated by the fructification organs (Apothecium or Perithecium). These spores are carried by the wind and taken to susceptible tissues of the plant, generally, some parts of the flower.

Sclerotiums have many shapes and during thresh can be easily included with the rest of the seed. Some examples of this are *Sclerotinia sclerotiorum* in horticultural, grain and flower crops, *Sclerotinia cepivorum* in garlic and *Claviceps purpurea* in barley.

Mix with infected Plant Parts

Infected plant's parts and residues can carry fructifications or spores of fungi and bacteria. This situation is particularly common in cereal, forage and oily crop's seeds. Examples of this are *Septoria nodorum* in wheat, *Puccinia malvacearum* in hollyhock (malvarrosa), *Colletotrichum trifolii* in clover, *Erwinia carotovora* pv. *carotovora* in tobacco, *Pseudomonas syringae* pv. *phaseolicola* in bean and *Sclerotinia sclerotiorum* in sunflower, soy and peanut.

Soil

Seeds can be mixed with contaminated soil and, carry micro-sclerotiums of *Macrophomina phaseolina* in dirty seeds of kidney bean, *Verticillium oxysporum* f.sp. *phaseoli* in cotton seeds, *Plasmodiophora brassicae* in turnip and *Fusarium oxysporum* f. sp. *phaseoli* in bean.

SEED TRANSMITION

We saw how the seed gets infected or infested. Now we will see how these seeds can produce an ill plant. This is to say, how the inoculum goes from the seed to the plant. We call seed-borne pathogens only those that can produce an infection. Because they have to be distinguished from those that can associate to the seed but do not produce an infection, these ones are called pathogens not transmitted by seed.

The virus that causes curly top in sugar beet can be present in the perisperm but does not cause an infection in the plant. In general terms, infection can be classified into systemic and non systemic. It is systemic when the pathogen introduces itself to the plant when the seed germinates, and develops with it. Non systemic infection occurs when there is a localized infection caused by the pathogen in the seedling at the stage of pre or post emergency, in this situation there are no systemic symptoms.

Systemic Transmition

This type of infection can be produced by pathogens that are carried with the seed in various parts, like the embryo, endosperm or episperm, or by contamination of the outer portion of this one.

Infection of the Embryo

When the seed germinates, if the embryo is infected, the pathogen initiates its growth together with the plant. Symptoms can show up during different stages of development. In the case of *Ustilago triticci*, the mycelium grows together with the plant and expresses symptoms only at the flowering stage, which are that all the tissues of the spike, exept the rachis, are replaced by spores.

Some pathovares of *Xanthomonas campestris* that infect cabbage, bean or sweet pepper move between the cells of the host until they reach the vascular system and produce symptoms in leaves or stems. Most of seed-borne viruses persist inside the embryo. Its multiplication and movement accompanies the plant during its development, and there can be symptoms at any stage, from the formation of first leaves until flowering or fructification.

Non Embryo Infection

Infection of the episperm (testa and endopleura) occasionally conducts to a systemic infection. Some bacteria, like *Corynebacterium michiganense* pv. *michiganense* in tomato and *Xanthomonas campestris* pv. *campestris* in cabbage, penetrate through stomata of cotyledons, and from there, reach the vascular system, initiating the systemic infection.

Episperm Contamination

In some few cases this type of contamination conducts to a systemic infection. Most of these exceptions are fungi that are highly specialized in their pathogenesis and produce in cereals the so called smuts, rusts or mildews.

In these cases, generally, spores are carried outside the seed, they germinate, penetrate the coleoptile, and start a systemic infection. This can occur directly o by a more complex system of haploid hypha fusion in genera like *Tilletia, Ustilago, Uromyces, Sclerospora, Pernospora and Puccinia.*

Non Systemic Transmition

Non systemic infection is very common, and in the same way as systemic infection, it can come from an infection, outside contamination or by pathogens mixed with the seeds.

Infection of the Embryo

This case is restricted to some pathogenic fungi that maintain itself in the embryo or the episperm as hypha inside the seed. Primary infection starts as injuries in the cotyledons or primary leaves, stems or petiole. Fungi fructifications (Pycnidium, Acervulus) can develop on these organs, under certain favorable conditions (temperature and humidity). These fructifications produce spores that, with the action of water and wind, disperse the disease to other parts of the plant and other plants. This occurs in *Ascochyta pisi* in pea, *Colletotrichum lindemuthianum* in bean and C. *truncatum* in soy.

Infection of the Episperm

Generally, seeds in which the episperm is infected, do not geminate, or germinate and contaminate the soil. Rarely produce a systemic infection, but can infect the seedling from the outside.

Over the injuries, new inoculum is produced; this one infects other parts of the plant or other plants. *Septoria nodorum* in wheat under high humidity conditions forms Picnidium in the coleoptile, whose spores disseminate the disease to other plants.

Contamination of the Episperm

Contamination outside the seed's testa can produce healthy seedlings, but the inoculum infects the soil and from there it can cause infections at more developed stages of the plant.

Wheat and rice grains that are contaminated with Teliospores of *Neovossia indica* in wheat and *N. torrida* in rice, produce Esporidisporas in the soil, that can be carried by the wind and infect flowers, originating the smuts.

Accompanying Contamination

The mixture of seeds with sclerotiums of the fungi or contaminated soil particles produces non systemic infections, at any stage of the plant growth and development. It is common for *Sclerotinia sclerotiorum* to produce mycelium from the sclerotiums.

These mycelium can directly infect the plant or produce fructifications that produce spores that can be carried by the wind and infect flowers of different species, like sunflower, soy, peanut, etc.

Prevention of Seedborne Diseases

Prevention of seed-borne diseases is based in two main facts. On one hand there is the usage of material free from pathogens and contaminants that could originate a disease and on the other there are the treatments to eliminate the inoculum in the seed or propagule.

When using material free from pathogens it is important to have adequate pathogen detection methods, this is to say, sensible and specific. No production of health quality seed can be done without valid methods to detect the possibility of contamination. Some treatments can be necessary to eliminate the pathogens from the seeds, or at leafs, maintain an economically viable balance.

Health Testing of the Seed

One of the most relevant aspects about controlled health quality seed is refereed to the detection methods used on every case. ISTA (International Seed Testing Association) appeared in 1924, by was only interested in seed's gene purity and germination aspects. Later then, thanks to the efforts made by Dr. L. C. Doyer publishing the first Manual for Determination of Seed Borne Disease and being the first director of the PDC (Plant Disease Committee), the importance of seed-borne diseases was turned unto first priority.

But detection methods were applied according to the criterion of each laboratory, so their results were not comparative. So, in 1957, PDC established a comparative health testing programme and standardize applied methods. The complexity for comparison of these methods made a considerable step back on the publication of working sheets (protocols). ISTA approved methods became Rules. By 1999 only 64 protocols were being compared, and only 14 of them were accepted as Rules.

As the solutions proposed by ISTA did not satisfy the needs of international seed industry, this industry, in 1994, organized The International Seed Health Initiative for Vegetables (ISHI-Veg) chartered by the vegetable seed industries in The Netherlands and France.

The Initiative was soon joined by the seed companies in the United States, Israel and Japan. This group represents the production of over 75% of the worlds vegetable seed supply. ISHI published many protocols, organized by crop. For horticultural crops, the ISHI-VEG, the Seed Health Testing Methods

ISHI-Veg Status

- An ISHI Validated Reference Method is a method that has been through the ISTA/ISHI comparative testing process and is being published as ISTA wn ISHI Reference Method is a method that has been through ISHI comparative testing and is under review by the ISTA/ISHI reviewers for publication as ISTA Working Sheet.
- An ISHI Accepted Method is a method that is commonly used by the seed industry for determining seed health. The method has been published in a journal, as a working sheet and/or is publicly available for comparative testing and commonly in use by the seed industry. The method is in comparative testing or aspects of the test are being tested for inclusion or deletion from the method.
- An ISHI Reviewed Method is a method that is publicly available or published, documented and prioritized by the ITG, but not subjected to a comparative test at this time. The method usually is accepted by other agencies or groups.

In 1995, during the second ISTA-PDC symposium it was decided that a Joint ISTA/ISHI Guidelines for Comparative Testing of Methods for Detection of Seed Borne Pathogens would be edited.

Through this combined effort many laboratories, organized under this alignments, the ISTA developed the Manual of Seed Health Testing Methods that has two sections:

- Validated Test Methods
- Peer Reviewed Methods.

The transition process from working protocol to ruled method is long and laborious, so working protocols can be considered as adequate methods for seed analysis. Nowadays, it is common that technological advances may become obsolete even before they are published.

The protocol to analyse *Xanthomonas campesrtis* pv.*campestris*, which causes black rot in crucifers, was published in 1982, there, it was recommended one minute to extract the bacteria from the seeds to starch culture medium. 2 years later, this time was raised to 2 hours, and other specific culture mediums. In 1987 the protocol was replaced, again in 1996 and at present time again it is under revision.

New Techniques for Seed Health Analysis

Pathogen detection methods have suffered great changes since 1957. Back then it was mainly targeted to fungi, and based exclusively on incubation and identification, or the cultivation of these seeds and counting the percentage of unhealthy seedlings. These analyses were very laborious and required a high infection level in order to be detected.

Nowadays, technological advances have allowed detecting even low infection levels, evaluating 1,000 to 10,000 seeds at once. Specificity of culture mediums, usage of antibiotics and other products has allowed isolating the pathogen from the seed. Together with this, the use of reactives as mono or polyclonal antiserums and molecular markers is also available. The Lettuce Mosaic Virus (LMV), back in 1950 the symptoms were observed over many thousands of seedlings from the seed that were going to be analyzed.

Later, technical advances determined that 30,000 seedlings were needed, and even later, in 1983, an alternative to the *Chenopodium quinoa* test was discovered, the ELISA test.

Watermelon Fruit Blotch caused by bacteria is a severe problem in glasshouse winter production. It was determined that a single contaminated seed in 9,000 of them was enough too widespread the bacteria to the rest of the plants. That is the reason why screening was done over 10,000 seedlings. Nowadays PCR technique is used.

Minimum requirements to Produce Certified Seed

The Organization for Economic Cooperation and Development (OECD) establishes steps to be taken in order to produce basic and certified seed. Between these steps it is emphasized that generations from the mother material to the basic material are strictly limited in number. The number of certified seed generations, after the basic seed, varies from crop to crop and technical conditions. In both, sexually and vegetative reproduced species, the schemes include similar steps. It generally starts with selection of a plant with genetic and health advantages. If health condition was not satisfying, certain actions should be taken in order to release it from considered pathogens. This Mother Plant is maintained under strict special security conditions in order to maintain healthiness (isolated glasshouses or chambers).

In the Certification Schemes two particularities abut seed borne pathogens should be considered, those transmitted exclusively with the seed and those that can also be transmitted with contaminated soil and plant rests. Diseases transmitted exclusively with the seed can be controlled by using clean healthy seed. This seed can be obtained by Seed Certification Programmes or by treatments that are really effective, like *in vitro* culture, thermotherapy, chemotherapy or a combined action of these, for systemic pathogens.

Tolerance levels in certified seed depend on the multiplication speed and dispersion of the inoculum of the pathogen of the considered disease. If a small amount of inoculum may be epidemic, like Anthracnose and the bean bacterial blight, if levels above zero chemical control may not be enough to stop the disease.

But pathogens whose inoculum multiplies slowly generation after generation, the disease may be evident only several generations later, so, if contamination levels are above zero may be accepted in adequate control methods are applied to the seed. This is what happens in wheat, with the Loose Smut, continuous treatments diminish inoculum levels, and make it perfectly tolerable.

PLANT DISEASE RESISTANCE

Plant disease resistance is crucial to the reliable production of food, and it provides significant reductions in agricultural use of fuel, land, water and other inputs. There are numerous examples of devastating plant disease impacts, as well as recurrent severe plant disease issues. However, disease control measures are reasonably successful for most crops. Across large regions and many crop species, it is estimated that diseases typically reduce plant yields by 10 per cent every year in more developed settings, but yield loss to diseases often exceeds 20 per cent in less developed settings.

Plant disease resistance derives both from pre-formed defences and from infection-induced responses mediated by the plant immune system. Relative to a disease-susceptible plant, disease resistance is often defined as reduction of pathogen growth on or in the plant, while the term disease tolerance describes plants that exhibit less disease damage despite similar levels of pathogen growth. Disease outcome is determined by the three-way interaction of the pathogen, the plant, and the environmental conditions (an interaction known as the disease triangle). Defence-activating compounds can move cell-to-cell and systemically through the plant vascular system, but plants do not have circulating immune cells so most cell types in plants retain the capacity to express a broad suite of anti-microbial defences.

Although obvious *qualitative* differences in disease resistance can be observed when some plants are compared (allowing classification as 'resistant' or 'susceptible' after infection by the same pathogen strain at similar pathogen inoculum levels in similar environments), a gradation of *quantitative* differences in disease resistance is more typically observed between plant lines or genotypes. Plants are almost always resistant to certain pathogens but susceptible to other pathogens; resistance is usually pathogen species-specific or pathogen strain-specific. Note that plant defence against herbivory (plant resistance to insect pests) exhibits some mechanistic similarities to, but also differences from, plant disease resistance (plant resistance to microscopic organisms).

Common Mechanisms of Plant Disease Resistance Pre-formed structures and compounds that contribute to resistance Plant cuticle/surface Plant cell walls Antimicrobial chemicals (for example: glucosides, saponins) Antimicrobial proteins Enzyme inhibitors Detoxifying enzymes that break down pathogen-derived toxins Receptors that perceive pathogen presence and activate inducible plant defences Inducible plant defences that are generated after infection Cell wall reinforcement (callose, lignin, suberin, cell wall proteins) Antimicrobial chemicals (including reactive oxygen species such as hydrogen peroxide, or peroxynitrite, or more complex phytoalexins such as genistein or camalexin) Antimicrobial proteins such as defensins, thionins, or PR-1 Antimicrobial enzymes such as chitinases, beta-glucanases, or peroxidases Hypersensitive response - a rapid host cell death response associated with defence mediated by 'Resistance genes.'

Plant Immune Systems and Plant Defence Signal Transduction Plant immune systems show some mechanistic similarities and apparent common origin with the immune systems of insects and mammals, but also exhibit many plant-specific characteristics. As in most cellular responses to the environment, defences are activated when receptor proteins directly or indirectly detect pathogen presence and trigger ion channel gating, oxidative burst, cellular redox changes, protein kinase cascades, and/or other responses that either directly activate cellular changes (such as cell wall reinforcement), or activate changes in gene expression that then elevate plant defence responses. Plants, like animals, have a basal immune system that includes a small number of pattern recognition receptors that are specific for broadly conserved microbe-associated molecular patterns (MAMPs, also called pathogen-associated molecular patterns or PAMPs). Examples of these microbial compounds that elicit plant basal defence include bacterial flagellin or lipopolysaccharides, or fungal chitin.

The defences induced by MAMP perception are sufficient to repel most potentially pathogenic microorganisms. However, pathogens express effector proteins that are adapted to allow them to infect certain plant species; these effectors often enhance pathogen virulence by suppressing basal host defences. Importantly, plants have evolved R genes (resistance genes) whose products allow recognition of specific pathogen effectors, either through direct binding of the effector or by recognition of the alteration that the effector has caused to a host protein. R gene products control a broad set of disease resistance responses whose induction is often sufficiently rapid and strong to stop adapted pathogens from further growth or spread. Plant genomes each contain a few hundred apparent R genes, and the R genes studied to date usually confer specificity for particular strains of a pathogen species.

As first noted by Harold Flor in the mid-20th century in his formulation of the gene-for-gene relationship, the plant R gene and the pathogen 'avirulence gene' (effector gene) must have matched specificity for that R gene to confer resistance. The presence of an R gene can place significant selective pressure on the pathogen to alter or delete the corresponding avirulence/

effector gene. Some R genes show evidence of high stability over millions of years while other R genes, especially those that occur in small clusters of similar genes, can evolve new pathogen specificities over much shorter time periods.

The use of receptors carrying leucine-rich repeat (LRR) pathogen recognition specificity domains is common to plant, insect, jawless vertebrate and mammal immune systems, as is the presence of Toll/Interleukin receptor (TIR) domains in many of these receptors, and the expression of defensins, thionins, oxidative burst and other defence responses. Some of the key endogenous chemical mediators of plant defence signal transduction include salicylic acid, jasmonic acid or jasmonate, ethylene, reactive oxygen species, and nitric oxide. Numerous genes and/or proteins have been identified that mediate plant defence signal transduction. Cytoskeleton and vesicle trafficking dynamics help to target plant defence responses asymmetrically within plant cells, towards the point of pathogen attack. Plant immune systems can also respond to an initial infection in one part of the plant by physiologically elevating the capacity for a successful defence response in other parts of the plant.

These responses include systemic acquired resistance, largely mediated by salicylic acid-dependent pathways, and induced systemic resistance, largely mediated by jasmonic acid-dependent pathways. Against viruses, plants often induce pathogen-specific gene silencing mechanisms mediated by RNA interference. These are primitive forms of adaptive immunity. In a small number of cases, plant genes have been identified that are broadly effective against an entire pathogen species (against a microbial species that is pathogenic on other genotypes of that host species). Examples include barley MLO against powdery mildew, wheat Lr34 against leaf rust, and wheat Yr36 against stripe rust.

An array of mechanisms for this type of resistance may exist depending on the particular gene and plant-pathogen combination. Other reasons for effective plant immunity can include a relatively complete lack of coadaptation (the pathogen and/or plant lack multiple mechanisms needed for colonization and growth within that host species), or a particularly effective suite of pre-formed defences (see above). Plant Breeding for Disease Resistance Plant breeders focus a significant part of their effort on selection and development of disease-resistant plant lines.

Plant diseases can also be partially controlled by use of pesticides, and by cultivation practices such as crop rotation, tillage, planting density, purchase of disease-free seeds and cleaning of equipment, but plant varieties with inherent (genetically determined) disease resistance are generally the first choice for disease control. Breeding for disease resistance has been underway since plants were first domesticated, but it requires continual effort. This is because pathogen populations are often under natural selection for increased virulence, new pathogens can be introduced to an area, cultivation methods can favour increased disease incidence over time, changes in cultivation practice can favour new diseases, and plant breeding for other traits can disrupt the disease resistance that was present in older plant varieties. A plant line with acceptable disease resistance against one pathogen may still lack resistance against other pathogens.

PLANT BREEDING FOR DISEASE RESISTANCE TYPICALLY INCLUDES

Identification of resistant breeding sources (plants that may be less desirable in other ways, but which carry a useful disease resistance trait). Ancient plant varieties and wild relatives are very important to preserve because they are the most common sources of enhanced plant disease resistance. Crossing of a desirable but disease-susceptible plant variety to another variety that is a source of resistance, to generate plant populations that mix and segregate for the traits of the parents. Growth of the breeding populations in a disease-conducive setting. This may require artificial inoculation of pathogen onto the plant population. Careful attention must be paid to the types of pathogen isolates that are present, as there can be significant variation the effectiveness of resistance against different isolates of the same pathogen species. Selection of disease-resistant individuals. Breeders are trying to sustain or improve numerous other plant traits related to plant yield and quality, including other disease resistance traits, while they are breeding for improved resistance to any particular pathogen.

Each of the above steps can be difficult to successfully accomplish, and many highly refined methods in plant breeding and plant pathology are used to increase the effectiveness and reduce the cost of resistance breeding. Resistance is termed *durable* if it continues to be effective over multiple years of widespread use, but some resistance 'breaks down' as pathogen populations evolve to overcome or escape the resistance. Resistance that is specific to certain races or strains of a pathogen species is often controlled by single R genes and can be less durable; *broad-spectrum resistance* against an entire pathogen species is often quantitative and only incompletely effective, but more durable, and is often controlled by many genes that segregate in breeding populations. However, there are numerous exceptions to the above generalized trends, which were given the names vertical resistance and horizontal resistance, respectively, by J.E. Vanderplank. Crops such as potato, apple, banana and sugarcane are often propagated by vegetative reproduction to preserve highly desirable plant varieties, because for these species, outcrossing seriously disrupts the preferred plant varieties. See also asexual propagation. Vegetatively propagated crops may be among the best targets for resistance improvement by the biotechnology method of plant transformation to add individual genes that improve disease resistance without causing large genetic disruption of the preferred plant varieties. Host Range There are thousands of species of plant pathogenic microorganisms (Plant Pathology), but only a small minority of these pathogens have the capacity to infect a broad range of plant species. Most pathogens instead exhibit a high degree of host-specificity. Non-host plant species are often said to express *non-host resistance*. The term *host resistance* is used when a pathogen species can be pathogenic on the host species but certain strains of that plant species resist certain strains of the pathogen species. There can be overlap in the causes of host resistance and non-host resistance. Pathogen host range can change quite suddenly if, for example, the capacity to synthesize a host-specific toxin or effector is gained by gene shuffling/mutation, or by horizontal gene transfer from a related or relatively unrelated organism. Epidemics and Population Biology Plants in native populations are often characterized by substantial genotype diversity and dispersed populations (growth in a mixture with many other plant species).

They also have undergone millions of years of plant-pathogen coevolution. Hence as long as novel pathogens are not introduced from other parts of the globe, natural plant populations generally exhibit only a low incidence of severe disease epidemics. In agricultural systems, humans often cultivate single plant species at high density, with numerous fields of that species in a region, and with significantly reduced genetic diversity both within fields and between fields. In addition, rapid travel of people and cargo across large distances increases the risk of introducing pathogens against which the plant has not been selected for resistance.

These factors make modern agriculture particularly prone to disease epidemics. Common solutions to this problem include constant breeding for disease resistance, use of pesticides to suppress recurrent potential epidemics, use of border inspections and plant import restrictions, maintenance of significant genetic diversity within the crop gene pool, and constant surveillance for disease problems to facilitate early initiation of appropriate responses. Some pathogen species are known to have a much greater capacity to overcome plant disease resistance than others, often because of their ability to evolve rapidly and to disperse broadly. Disease resistance in fruit and vegetables There are a number of lines of defence against pests (that is, those animals that cause damage to the plants we grow) and diseases in the organic garden, principal among these being the practice of good husbandry, creating healthy soil and ensuring high standards of garden hygiene. But no matter how diverse and healthy the garden eco-system may be, there will always be a degree of disease and pest presence.

In many ways, some level of pathogen population in the garden can be not only acceptable but desirable as they are indicative of a generally healthful and diverse environment, and add to the overall robustness of the system as an immunity to such detrimental influences will build up, particularly in a balanced polycultural regime. Indeed, most of the plants we grow will tend to be selected because they are trouble free, and those that are more susceptible to attack will have fallen by the wayside over time. However, most farmers find it unacceptable that the food crops they grow are damaged by pests.

For these crops there has been considerable research and selective breeding carried out in order to find cultivars that are resistant or immune to pest and disease damage. Breeding for plant disease resistance generally has involved finding suitable genetic material amongst existing stocks or in the wild, which is then incorporated into commercial varieties. Example: The apple In the case of apples, in which research is being carried out in order to develop resistance to diseases such as apple scab (*Venturia inaequalis*), powdery mildew (*Podosphaera leucotricha*), orchard fireblight (*Erwinia amylovora*), woolly apple aphid (*Eriosoma lanigerum*) and collar rot (*Phytophthora cactorum*), the main sources of resistant material used in breeding programmes such as those being run by East Malling Research Station in England or Hortresearch in New Zealand are major gene resistances derived from crab-apples.

The Vf gene for black spot resistance is derived from the ornamental crab-apple species *Malus floribunda*. Most black spot resistant cultivars developed around the world carry this gene, but there are some selections that carry the Vr (from M. pumila) or Vm (from M. micromalus) gene. Major gene resistances to powdery mildew are derived from M. robusta (Pl1) and M. zumi (Pl2), and the apple cultivar *Northern Spy* has a long-standing reputation for its major gene resistance to woolly apple aphid. Since early this century this resistance has been used to develop woolly aphid resistant rootstocks such as MM.106 and M.793. Much later it was shown that the cultivar was also very resistant to collar rot and a useful breeding parent for this resistance. Resistance and immunity Some plants can tolerate the presence of large numbers of insects without being severely affected.

This is not very satisfactory however as insects will still cause damage, and in fact further breeding and population expansion of the pest species is supported. Other varieties are less attractive to pests, but this can be difficult to sustain or demonstrate.

The most valuable form of resistance is where the pest cannot survive as well on one variety as on another. In some cases this can actually make the plants immune to attack, as is the case with the lettuces *Avoncrisp* and *Avondefiance* which were bred at the Institute of Horticultural Research, Wellesbourne during the 1960s, which are fully resistant to lettuce root aphid (*Pemphigus bursarius*). Trade-off of breeding for resistance Sometimes however there can be a *trade-off*, for those varieties which have increased immunity or resistance may be lacking in other qualities such as flavour, yield or quality. Celery resistant to the Fusarium fungus (*Fusarium oxysporum spp.*) may not succumb to this disease, but may also be unacceptably short, ribby and low yielding.

Further, a cultivar that is resistant to one disease may be more susceptible to another that is equally important. A lettuce cultivar that is resistant to mosaic virus may be sensitive to corky root disease, whilst another that resists corky root may be vulnerable to downy mildew (*Brim lactic*).

Another drawback to resistance is that depending on the host pathogen system, resistance is sometimes not long lasting as new pathogen strains quickly develop, and further research and breeding is constantly needed. Availability of resistant varieties Resistant varieties are not available for all crops. For several of the most damaging plant diseases, such as Potato blight (*Phytophthora infestans*) and white rot (*Sclerotic cepivorum*) of the Allium family, no acceptable resistant cultivars are yet available.

DISEASE MANAGEMENT IN SEED CROPS

In the Seeds of Change Quality Control Programme we work with our seed growers to avoid disease in seed production, we carefully inspect and clean our seed at our own certified organic seed cleaning facility, and we test many of our seeds for specific diseases of concern. One of the most important factors in avoiding disease in seed crops is producing in regions with minimal disease pressure. Many of the Seeds of Change growers are located in western regions where dry summers aid in minimizing disease pressure and allow harvesting seed under dry, late-summer/fall conditions. Additionally we work with our growers and professional breeders to select several crops for disease resistance.

Careful field management in seed production can significantly mitigate seed-borne diseases in seed crops. Many of the same practices that prevent disease in field crops are used to prevent disease in seed crops. Disease promoting conditions are avoided by not using overhead irrigation under moist conditions, using drip irrigation, and cutting water to allow seed crops to dry before harvesting. Spacing plants adequately and orienting rows with the prevailing winds aids in increasing air flow—an important factor in minimizing disease. Rotating crops is important to avoid disease build-up in the soil. Of course, starting with clean, disease-free seed for planting stock is also crucial. Additionally, seed growers must use care in harvesting and cleaning seed to avoid post-harvest infection.

Seeds of Change works with our seed growers to select varieties for disease resistance. In this programme disease is intentionally allowed to build up in an isolated field for breeding and selection purposes. Seed crops are selected under the disease conditions for resistance to the disease. In this case care is taken to ensure the stock seed is not harboring seed-borne diseases when used as planting stock for seed production. Many of Seeds of Change beet varieties have been selected in this manner for resistance to Rhizoctonia, a bacteria that causes scarring and lesions on beet roots. Many of the lettuce varieties offered have also been selected for resistance to Downy Mildew and Sclerotinia. Seeds of Change growers regularly practice rouging (selection) in seed production fields by removing diseased plants from the field. Over 1,500 microorganisms have been shown to be related to seeds of all types. However only a small percentage of these organisms pose a threat of disease to the following crop. The majority of common diseases are not seed-borne, but initiate from the local environment. As most diseases are regionally based it is valuable to check with local extension services and growers to identify which seed-borne diseases pose a threat in your ecosystem and then place extra care in sourcing varieties of these crops. If there are crops of particular concern, ask your seed company for recommendations and information about their disease management programme. And always use good organic disease management practices in the field.

CAUSE

Black rot is caused by a bacteria, *Xanthomonas campestris* pv. *campestris,* that can infect most crucifer crops at any growth stage. This disease is difficult for growers to manage and is considered the most serious disease of crucifer crops worldwide.

The disease can cause significant yield losses when warm, humid conditions follow periods of rainy weather during early crop development. Late infections can provide a wound for other rot organisms to enter and cause significant damage during storage.

Fig. Black rot Infected Cabbage.

SYMPTOMS

Symptoms of black rot vary considerably depending on the host, cultivar, plant age and environmental conditions. The bacteria can enter plants through natural openings and wounds caused by mechanical injury on roots and leaves. Seedborne bacteria infect the emerging seedlings through pores on the margin of the cotyledons and then spread systemically through the seedling. Infected seedlings grown in the greenhouse under cool conditions (below 15–18 °C) frequently do not show any symptoms of the disease. When infected seedlings are transplanted to the field and temperatures rise to 25–

35 °C during periods of high relative humidity (80–100%), they become stunted with dead spots on the cotyledons and will eventually wilt, and die. In regions with temperate climates (where temperatures remain cool), disease symptoms on infected seedlings may not always be obvious or appear severe. Infected seedlings grown under cool conditions may ooze bacteria from pores and lesions, which then serve as a source of the pathogen for neighbouring plants.

Fig. Young Cabbage leaf with V-shaped Lesion Characteristic of black rot Symptoms.

On older plants, the disease symptoms often appear as yellow or dead tissue at the edges of leaves, similar to tip burn, except the lesion frequently progress into a V-shape with the base of the V usually directed along a vein. Close inspection of infected leaves and stems may reveal black veins running through the infected tissue from which the disease gets its name. Lesions on leaves can expand down towards the base of the leaf causing the leaf to wilt and die.

Fig. Black rot Symptoms Appear as dead Tissue at the tips of (a) Kale, (b) Cauliflower, and (c) Cabbage Leaves. Note the V-Shaped Lesion Progressing from the tip along the vein of the black rot Infected Cabbage leaf.

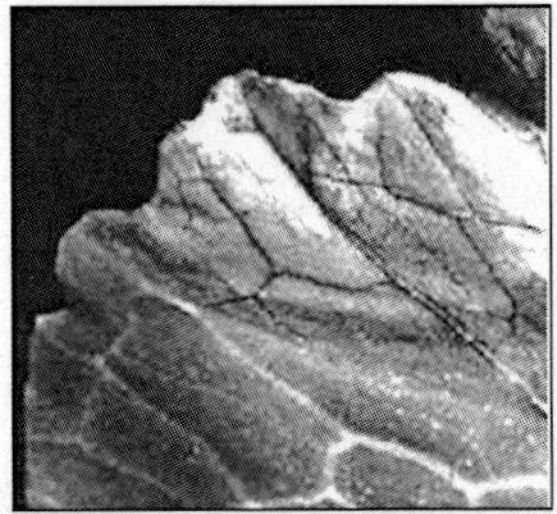

Fig. Black Rotting Veins Running through Black Rot Lesion on tip of Cauliflower Leaf.

The bacteria produce a sticky polysaccharide called xanthan that eventually plugs the vascular tissue inside the veins causing them to collapse and turn black. The tissue above the plugged, collapsed xylem eventually turns yellow, wilts and dies. During hot humid environmental conditions, the bacteria can move from the leaf into the stem through the xylem.

Once inside the stem, the bacteria can move up or down to other parts of the plant including the roots. Systemically infected plants may produce chlorotic areas anywhere on the leaf. Severely infected leafy cole crops such as kale and cauliflower tend to shed their leaves from the bottom up leaving only a tuft of distorted leaves separated from the root system by a scarred barren stem. Symptoms on cauliflower often appear as black flecks or scorched leaf margins. The curds of infected cauliflower heads often become blackened. Foliar symptoms may not be visible on infected root crops such as rutabaga and radish but blackened vascular tissue can appear inside the edible root tissue rendering the plants unmarketable. Although some infected plants may appear healthy, cutting across infected stems will reveal characteristic blackened vascular tissue. This is a simple method of determining the presence of the disease.

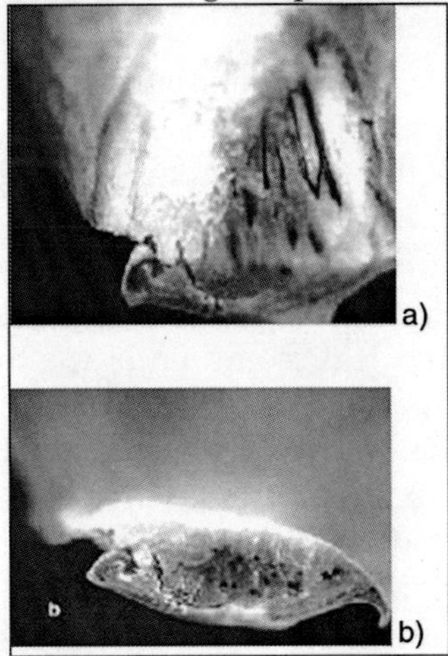

Fig. Cross Section of the base of a Black rot Infected Cabbage (a) Stem and (b) Leaf Revealing the Black, Collapsed Xylem.

Some symptoms of black rot closely resemble those caused by *Fusarium* yellows, which causes the vascular tissue to turn brown. Most commercial crucifer cultivars are resistant to *Fusarium*.

Fig. Leaf Symptoms of Fusarium Yellows Sometimes Appear Similar to black rot Except the Vascular Tissue turns brown Instead of black.

DISEASE SPREAD

Seed contaminated with black rot bacteria is considered the most important source of the pathogen and significantly contributes to the spread of this disease worldwide. As few as 3 infected seeds per 10,000 (0.03% infected seeds) can result in a black rot epidemic. Seed should be tested and certified to be disease free with less than 1 in 30,000 infected seed.

The organism survives in infected crop tissue left on the soil until the crop tissue rots. However, the bacteria do not survive very long in soil as unprotected free living organisms. The black rot bacteria can also infect and survive on many crucifer weeds. This also contributes to the persistence and spread of the disease. It can grow and multiply on host tissue without infecting or causing disease.

Rain splashed bacteria from contaminated plant residue left on the soil or from neighbouring diseased plants is the primary method of disease spread throughout a field. The bacteria enter and exit through water-secreting glands called hydathodes located at the edges and tips of leaves. Hydathodes often produce a drop of water during periods of high humidity early in the morning. The pathogen spreads very quickly when rain droplets contaminated with bacteria splash onto healthy leaves and enter the hydathodes.

The bacteria move into the leaf veins through hydathodes and begin to multiply, rot and plug the veins. Contaminated water droplets that exude out of hydathodes of infected leaves can then be rain- splashed to other plants. Black rot is more severe and widespread in fields that receive frequent early morning rains, particularly in May and June.

Equipment, people, animals and overhead irrigation can further spread the disease. Insects can also spread the bacteria; however, their contribution to the spread of black rot is limited.

Fig. Hydathodes are Special Glands or Pores at the end of Vascular Tissue on leaves through which Water Exudes and are a Natural Opening for Black Rot Bacteria to Infect.

DISEASE MANAGEMENT

Black rot management begins with the identification of potential disease sources and utilising an Integrated Pest Management (IPM) strategy including host resistance, planting disease free seed, avoiding spreading the disease and proper sanitation. Sanitation is the main method that reduces, excludes or eliminates the initial sources of disease. General sanitation practices include crop rotation, disinfecting seed, rouging diseased plants, elimination of refuse piles and eradication of alternative hosts.

SEED TREATMENT

Seedborne inoculum significantly contributes to the spread of black rot bacteria. Growers should only plant tested certified seed < 1 infected seed in 30,000 or 0.003% contamination. When the infection level of seed is not known or disease-free seed is not available, seed should be treated to eliminate the bacteria. Growers who purchase transplants should request proof the seedlings were grown from disease-free or treated seed. During transplanting, diseased seedlings should not be planted in the field.

Seed treatments do not always eliminate 100% of the bacteria on or in the seed, and may adversely affect seed germination and vigour. Soaking seeds in hot water at 50 °C for 25–30 min. is the most effective treatment for seedborne blackrot control. Weak seed, seed stored for several years and seed of certain crucifer crops; such as, cauliflower, kohlrabi, kale, rutabaga and summer turnip, may be damaged by hot water treatment; soak for 15 min. at 50 °C only.

The effect of the hot water seed treatments on every variety of each individual crucifer crop has not been investigated. Growers are encouraged to treat a small portion of seed and plant in pots to determine the effect of the seed treatment on germination and vigour, prior to treating the entire seed lot.

AVOID DISEASE SPREAD

Use new seed trays each year to avoid contaminating this year's crop with residual black rot bacteria from the previous year. If purchasing new trays each year is not economically feasible, used trays can be sterilised with steam, boiling water or chemical disinfectants to eliminate potential contamination. Destroy infected seed trays immediately to prevent disease spread to other seedling trays.

Avoid soaking crates or bundles of transplant seedlings in tubs of water before transplanting. The black rot bacteria can spread from diseased to healthy seedlings by infecting leaf scars and wounds on roots when soaked in water.

Black rot bacteria can contaminate the surface of clothing, equipment, tools and water sources. Reducing seeding rates and densities to promote good air circulation, facilitating the quick drying of plants, timing irrigation when plants will dry quickly and restricting field activities until later in the day when fields are dry will help reduce disease spread. Working in diseased fields last will also avoid disease spread from infected to non-infected fields. Wash and disinfect equipment before moving from one field to another.

FIELD SELECTION

Field selection is very important due to the distance the pathogen can spread. Whenever possible, select fields as far away from fields grown to crucifer crops the previous year. Select fields that are well drained and will not receive run-off water from areas or fields where crucifers have been grown previously. Well drained, light soils are best for crucifer production because they can be worked early in the season and facilitate earlier planting of transplants. Planting early can help avoid disease because environmental conditions are usually not conducive for the development and spread of black rot bacteria.

6

The Botany of Seeds

Seeds provide the vital genetic link and dispersal agent between successive generations of plants. Angiosperm seeds are produced and packaged in botanical structures called fruits which develop from the "female" pistils of flowers. Immature seeds (called ovules) each contain a minute, single-celled egg enclosed within a 7-celled embryo sac. The haploid egg is fertilized by a haploid sperm resulting in a diploid zygote that divides by mitosis into a minute, multicellular embryo within the developing seed. A second sperm unites with 2 haploid polar nuclei inside a binucleate cell called the endosperm mother cell which divides into a mass of nutritive tissue inside the seed.

In most seeds the embryo is embedded in this endosperm tissue which provides sustenance to the embryo during germination. In exalbuminous seeds (found in many plants such as the legumes), the endosperm tissue is already absorbed by the time you examine a mature seed within the pod, and the 2 white fleshy halves in the seed are really the cotyledons (components of the embryo).

The 2 sperm involved in the double fertilization process originated within the pollen tube that penetrated the embryo sac. The pollen grain (and pollen tube) come from the "male" organs (called anthers) on the same plant or different parental plants in a remarkable process known as pollination. Pollination is also accomplished by the wind (or water), and it may also involve insects in some of nature's most fascinating relationships between a plant and an animal. This is especially true of the amazing fig trees and their symbiotic wasps.

One of the important functions of seeds and fruits is dispersal; a mechanism to establish the embryo bearing seeds in a suitable place away from their parental plants. There are 3 main mechanisms for seed and fruit dispersal:

1. Hitchhiking on animals,
2. Drifting in ocean or fresh water, and
3. Floating in the wind. This article concerns one of the most remarkable of all seed dispersal methods, riding the wind and air currents of the world.

Gliders

Gliders include seeds with 2 lateral wings that resemble the wings of an airplane. They become airborne when released from their fruit and sail through the air like a true glider. One of the best examples of this method is Alsomitra macrocarpa, a tropical vine in the Gourd Family (Cucurbitaceae) native to the Sunda Islands of the Malay Archipelago. Football-sized gourds hang from the vine high in the forest canopy, each packed with hundreds of winged seeds. The seeds have two papery, membranous wings, with combined wingspans of up to 5 inches (13 cm). They reportedly inspired the wing design of some early aircraft, gliders and kites. Although the seeds vary in shape, some of the most symmetrical ones superficially resemble the shape of the "flying wing" aircraft or a modern Stealth Bomber. the aerodynamic seeds spiral downward in 20 foot (6 meter) circles, although a gust of wind would probably carry them much farther away.

Parachutes

An individual parachute of western salsify (Tragopogon dubius) showing an umbrella-like, plumose crown of hairs (pappus) above a slender one-seeded fruit (called an achene).

These fragile units can become airborne with the slightest gust of wind, and can literally sail across valleys and over mountain slopes.Western salsify or goatsbeard (Tragopogon dubius) showing dense, puff-like cluster of numerous parachute seeds (one-seeded achenes). Each achene has an umbrella-like crown of plumose hairs and may literally be carried into the atmosphere by strong ascending air currents. A population explosion of western salsify(Tragopogon dubius) near Mono Lake, on the east side of the Sierra Nevada of Central California. This ubiquitous species is actually native to Europe and Asia.

Parachutes include seeds or achenes (one-seeded fruits) with an elevated, umbrella-like crown of intricately-branched hairs at the top, often produced in globose heads or puff-like clusters. The slightest gust of wind catches the elaborate crown of plumose hairs, raising and propelling the seed into the air like a parachute. This is the classic mechanism of dispersal for the Eurasian dandelion (Taraxacum officinale) and includes numerous weedy and native members of the Sunflower Family (Asteraceae). A giant Eurasian version of the dandelion called salsify or goat's beard (Tragopogon dubius), is one of the most successful wind-travelers in North America. Its seeds have literally blown across mountain ranges, colonizing vast fields of open land in the western United States.

Three weedy species of salsify (T. dubius, T. pratensis and T. porrifolius) have been introduced into the western United States, 2 with yellow dandelion-type flowers and one with purple flowers. The latter, purple-flowered species (T. porrifolius) has a large, edible tap root with a flavor resembling oysters, hence the name "oyster plant. Inflorescence and mature, seed-bearing head of the Eurasian dandelion (Taraxacum officinale). The slightest gust of wind catches the elaborate crown of plumose hairs, raising and propelling each seed-bearing achene into the air like a parachute. This successful weed thrives in a wide range of climates and has become naturalized throughout North America

In some parachutes, the crown of silky hairs arises directly from the top of the seed (not on an umbrella-like stalk). Again, the Sunflower Family (world's largest plant family with about 24,000 described species) contains many weedy representatives with this type of parachute seed. One of the most troublesome weeds of farm land in the western United States is wild or thistle artichoke (Cynara cardunculus). The large seed head of this weedy composite releases hundreds of parachute seeds which fly through the air and invade vast areas of grazing land with spiny, perennial bushes that literally take over. The large leaf stalks (resembling giant celery stalks) are edible and are sold under the name of "cardoon." Populations of wild artichoke often contain so much variation between spiny and non-spiny plants, that some experts believe that they belong to one variable species. In fact, some botanists believe that the cultivated artichoke (C. scolymus) may be a cultivated variety of the wild C. cardunculus. Incidentally, the delicious artichoke is really a cooked flower head in which the outer bracts (phyllaries) and central basal portion (receptacle) are dipped in butter and eaten.

Brown puffs (Stebbinoseris heterocarpa), formerly Microseris heterocarpa, an interesting member of the sunflower family. A closely related species is called silver puffs (Uropappus lindleyi syn. Microseris lindleyi). In Stebbinoseris the pappus paleae are bifid at the apex. Unlike the weedy dandelions (Taraxacum) and salsify (Tragopogon), this is a native species in California.

Another plant family which has evolved this parachute method of seed dispersal is the Milkweed Family (Asclepiadaceae). Hundreds of parachute seeds (each with a tuft of silky hairs) are produced within large, inflated pods called follicles. So abundant are the silky hairs, that they were actually collected and used as a substitute for kapok during World War II. Kapok comes from masses of silky hairs that line the seed capsules of the kapok tree (Ceiba pentandra), an enormous rain forest tree of Central and South America. Kapok is used primarily as a waterproof filler for mattresses, pillows, upholstery, softballs, and especially for life preservers. The floss silk tree (Chorisia speciosa), another member of the Bombax Family (Bombaceae) also produces large seed capsules lined with masses of silky hairs. This tree with its distinctive thorny trunk and showy pink flowers is commonly planted in southern California. The seeds of kapok and floss silk trees are embedded in these silky masses which aid in their dispersal by wind; (Cottony Seeds & Fruits). The Dogbane Family (Apocynaceae) also includes members with seed pods (follicles) and parachute seeds similar to those of milkweeds. One of the best examples is Nerium oleander, a drought-resistant, Mediterranean shrub planted throughout southern California. The foliage contains a powerful cardiac glycoside that can permanently relax the heart muscle.Parachute seeds escaping from the follicle of Nerium oleander. The crown of silky hairs arises directly from the top of the seed (not on an umbrella-like stalk. Unlike the seed-bearing achenes of the sunflower family (Asteraceae), these are true seeds.

Helicopters (Whirlybirds)

The South American tipu tree (Tipuana tipu) has one of the most unusual legumes in the world. Unlike the fruits of most members of the legume family (Fabaceae), the 3rd largest plant family, the fruits of this tree have a distinctive wing that causes the legume to spin as it falls from the rain forest canopy.

Helicopters (also called Whirlybirds) include seeds or one-seeded fruits (samaras) with a rigid or membranous wing at one end. The wing typically has a slight pitch (like a propeller or fan blade), causing the seed to spin as it falls. Depending on the wind velocity and distance above the ground, helicopter seeds can be carried considerable distances away from the parent plant. The spinning action is similar to auto-rotation in helicopters, when a helicopter "slowly" descends after a power loss.

Numerous species of flowering trees and shrubs in many diverse and unrelated plant families have evolved this ingenious method of seed dispersal, good examples of convergent evolution. Representative examples of helicopter seeds and one-seeded fruits (called samaras) include the Maple Family (Aceraceae): Maples and box elder (Acer); Olive Family (Oleaceae): Ash (Fraxinus); Legume Family (Fabaceae): Tipu tree (Tipuana tipu); and the Protea Family (Proteaceae): Banksia and Hakea.

An interesting one-seeded winged fruit that spins as it falls through the air. It is called "whirling nut" and belongs to the genus Gyrocarpus in the gyrocarpus family (Gyrocarpaceae). Note: This genus of tropical trees, shrubs and lianas is often placed in the hernandia family (Hernandiaceae). The unusual fruit shown above was collected and photographed at Ho'omaluhia Botanical Garden on the windward side of Oahu in the Hawaiian Islands. Identification provided by Ricarda Riina, Botany Department, University of Wisconsin. Although they are classified as gymnosperms with naked seeds arising from woody cones rather than flowers, the Pine Family (Pinaceae) contains many genera with winged seeds, including Pinus (Pine), Abies (fir), Picea (spruce), Tsuga (hemlock), and many additional genera. When shed from cones high on upper branches, they fly over slopes and across deep canyons. The natural reforestation of conifers following fire is proof of the flying ability of seeds from nearby forested slopes.

Maples have a double or twin samara composed of 2 winged one-seeded fruits (double samara) joined together at their bases. When they break apart, each winged fruit flies like a typical helicopter seed. Although the Legume

Family (Fabaceae) is the third largest plant family with over 18,000 described species, the vast majority of legumes do not have winged seeds or fruits. The South American tipu tree (Tipuana tipu) is a notable exception, with beautiful yellow blossoms that give rise to pendant, samara-like legumes, each with a large wing on the lower end. The dried, winged legumes spin so neatly in the air that they could be marketed as a child's toy.

The remarkable Protea Family (Proteaceae) of Australia contains some truly amazing genera with winged seeds, including Banksia and Hakea. Although they are flowering plants, banksias produce a dense flower cluster (inflorescence) that gives rise to a cone-like structure containing many woody carpels. Each carpel bears 2 winged seeds and the entire cone-like structure superficially resembles a pine cone. In fact, some banksias release their seeds following fire and even resprout from subterranean lignotubers like chaparral shrubs. The native range of hopseed bush (Dodonea viscosa), a member of the Soapberry Family (Sapindaceae), extends from Arizona to South America. It is also commonly cultivated in southern California. The papery, winged fruits flutter and spin in the air, and may be carried short distances by the wind. The jacaranda tree (Jacaranda mimosifolia) of northwestern Argentina. Like many other members of the Bignonia Family (Bignoniaceae), the papery, winged seeds flutter and spin as they are carried by the wind.

Although their mode of dispersal is similar to single-winged helicopter seeds, the flutterer/spinners include seeds with a papery wing around the entire seed or at each end. When released from their seed capsules they flutter or spin through the air. Whether they spin or merely flutter depends on the size, shape and pitch of the wings, and the wind velocity. This method of wind dispersal is found in numerous species of flowering plants in many different plant families. Some examples of flutterer/spinner seeds include the Quassia Family (Simaroubaceae): Tree of heaven (Ailanthus altissima); Figwort Family (Scrophulariaceae): Empress tree (Paulownia tomentosa); Bignonia Family (Bignoniaceae): Jacaranda (Jacaranda mimosifolia), catalpa (Catalpa speciosa), desert willow (Chilopsis linearis), yellow bells (Tecoma stans), bower vine (Pandorea jasminoides), violet trumpet vine (Clytostoma callistegioides), and the fabulous trumpet trees (Tabebuia serratifolia and T. ipe); Elm Family (Ulmaceae): American and Chinese elms (Ulmus americana and U. parvifolia); Soapberry Family (Sapindaceae): Hop seed (Dodonea viscosa); and the Goosefoot Family (Chenopodiaceae): Four-wing saltbush (Atriplex canescens).

Any discussion of flutterer/spinners would not be complete without mentioning the quipo tree (Cavanillesia platanifolia), a massive rain forest tree in the bombax family (Bombacaeae) native to Panama. The enormous winged fruits of the quipo tree flutter through the air, carpeting the ground beneath the huge canopy of this striking tropical tree.The quipo tree (Cavanillesia platanifolia), a remarkable rain forest tree in the bombax family

(Bombacaceae) with huge winged fruits. This massive tree is native to Panama. Some of the most beautiful flowering trees of the New World tropics belong to the Bignonia Family (Bignoniaceae). They typically produce long, slender (cigar-shaped) seed capsules containing masses of flat seeds with papery wings at each end. [The beautiful jacaranda of Argentina has flattened, circular seed capsules.] The lovely yellow bells (Tecoma stans) is native to Mexico and the Caribbean region, and is the official flower of the U.S. Virgin Islands. Some of the South American trumpet trees, including the pink-flowered Tabebuia avellanedae and the yellow-flowered Tabebuia serratifolia, are also called ironwoods or axe-breakers (quebrachos) because of their dense, hard wood. The latter species is called "pau d'arco" and its wood actually sinks in water, with a specific gravity of 1.20. In South America, trumpet trees drop their leaves during the dry season and produce a profusion of pink or yellow blossoms. The crowns of these huge timber trees resemble gigantic floral bouquets in the midst of the forest. As with so many tropical species, some of the trumpet trees inhabit rain forest areas that are seriously threatened by slash and burn agriculture, large plantations of exportable products, and the general annihilation of the South American rain forests.

Cottony seeds and fruits include seeds and minute seed capsules with a tuft (coma) of cottony hairs at one end, or seeds embedded in a cottony mass. Some of the examples in this group are very similar in function to parachute seeds, but probably are not carried as far by the wind. Many plant families have this type of wind dispersal, including the Willow Family (Salicaceae): Willows (Salix) and Cottonwoods (Populus); Cattail Family (Typhaceae): Cattails (Typha); Evening Primrose Family (Onagraceae): Willow-Herb (Epilobium) and California fuchsia (Zauschneria); Bombax Family (Bombaceae): Kapok tree (Ceiba pentandra) and floss silk tree (Chorisia speciosa); and the Sycamore Family (Platanaceae): Sycamore (Platanus).

In the California sycamore (Platanus racemosa), a common riparian (streamside) tree throughout the state, the one-seeded fruits (achenes or nutlets) are produced in dense, globose heads. The spherical heads hang from branches like little balls. Individual achenes have a tuft of hairs at the base which probably helps in their wind dispersal. Seeds of the South American kapok tree (Ceiba pentandra) and floss silk tree (Chorisia speciosa) are embedded in dense masses of silky hairs inside large woody capsules. This undoubtedly helps to disperse the seeds when seed-bearing masses of hair are carried by the wind. In tropical regions of the New World, the kapok grows into an enormous rain forest tree with a massive buttressed trunk. Kapok hairs are coated with a highly water-resistant, waxy cutin layer. The empty lumen (cavity) inside each hair is larger the cotton hairs; hence, the hairs are lighter. Unlike cotton hairs, kapok is difficult to spin and is not made into textiles. It is used primarily as a waterproof filler for mattresses, pillows, upholstery, softballs, and especially for life preservers. A kapok-filled life jacket can support 30 times its own weight in water. One fuzzy brown cattail spike may contain a million tiny seeds.

Each seed has a tuft of silky white hairs and is small enough to pass through the "eye" of an ordinary sewing needle. They are shed in clouds of white fluff and float through the air like miniature parachutes. A cattail marsh covering one acre may produce a trillion seeds, more than 200 times the number of people in the world. The fluffy seeds have been used for waterproof insulation and the buoyant filling of life jackets. In addition, each plant produces billions of wind-borne pollen grains; in fact, so much pollen that it was used as flour by North American Indians and made into bread. Cottonwoods and willows also produce masses of seeds, each with a tuft of soft, white hairs. Since they are dioecious, with pollen-bearing male and seed-bearing female trees in the population, only female trees produce the actual cotton. During late spring and summer in the western United States, the cottony fluff from cottonwoods resembles newly fallen snow. Because the wind-blown fluff can be quite messy in cultivated parks and gardens, male trees are generally planted. The discriminatory label of "cottonless cottonwood" refers to a male tree.

Tumbleweed (Russian Thistle)

The common tumbleweed or Russian thistle is a rounded, bushy annual introduced into the western United States from the plains of southeastern Russia and western Siberia in the late 1800s. The name "thistle" comes from the stiff, sharp-pointed, awl-shaped leaves. Although it is depicted in songs of the old west, this species is a naturalized weed in North America. It is listed in most older references as Salsola kali or S. pestifer; however, the Jepson Flora of California (1993) lists it as S. tragus. Russian thistle belongs to the goosefoot family (Chenopodiaceae), along with many weedy species and some valuable vegetables, including beets (Beta vulgaris), goosefoot (Chenopodium album) and spinach (Spinacia oleracea).A large tumbleweed (Salsola tragus) in San Diego County, California. Tumbleweeds are pushed along by the wind, scattering thousands of seeds as they roll across open fields and valleys. A tumbleweed of this size is difficult to hold on to during a strong wind storm.

Tumbleweed is a prolific seeder and rapid seed germination and seedling establishment occurs after only a brief and limited rainy season. A single plant may produce 20,000 to 50,000 seeds within numerous small fruits, each surrounded by a circular, papery border. Mature plants readily break off at the ground level and are pushed along by strong gusts of wind. As they roll along hillsides and valleys, the seeds are scartered across the landscape. Tumbleweeds often pile up in wind rows along fences and buildings. This is a troublesome weed in agricultural areas because it literally covers the farm land with bushy, prickly shrubs. One interesting use for this plant in arid regions of the American southwest is for a "snowman" at Christmas time.

Three proportionally sized tumbleweeds are used to make the head, thorax and main body of a "snowman." Another suggested use is to compress tumbleweeds into logs and use them for firewood.

Miscellaneous

Squirrel-Tail Grass (Elymus elymoides), formerly named Sitanion hystrix is an attractive grass native to the mountains and plains of the western United States. Seed-bearing spikelets) of the flower spike (containing one-seeded fruits called grains and very long awns) are carried short distances by the wind. Although not as efficient fliers, the long awns function like the parachute bristles (pappus) of composites.

This miscellaneous category of wind-blown seeds and fruits includes plants that really don't fit the above 5 categories. The Grass Family (Poaceae) includes a number of species with plumose flower stalks that fragment into seed-bearing spikelets that blow into the wind. Some of these species have become troublesome weeds in southern California, including the South African fountain grass (Pennisetum setaceum). Although this tufted perennial makes an attractive, drought-resistant landscaping plant along walkways and roads, it is becoming a widespread weed in disturbed areas of San Diego County. Another species, called squirrel-tail grass (Elymus elymoides), resembles a weedy introduced grass, but it is actually a native perennial of dry, rocky mountains and open land in the western United States. To appreciate its airborne seeds, you really must see this grass during a strong gust of wind on the eastern slopes of the Sierra Nevada during late summer.

Mountain mahogany (Cercocarpus minutiflorus), a native shrub in the chaparral of southern California, produces a rather unique wind-blown fruit. The one-seeded fruit (achene) has a persistent, feathery style that glistens in the sunlight. Although they usually don't travel very far, the achenes are blown into the air by strong gusts of wind during the dry, fire season of late summer and fall. This species is not related to the West Indian mahogany (Swietenia mahagoni) or the Honduran mahogany (S. macrophylla), members of the true Mahogany Family (Meliaceae). Mountain mahogany actually belongs to the Rose Family (Rosaceae) and produces very hard wood that sinks in water when dry. In fact, the wood of a montane species (C. ledifolius), has a specific gravity of 1.12, as heavy and dense as ebony (Diospyros ebenum).

CHARLES DARWIN AND OCEAN DISPERSAL OF SEEDS

Plant dispersal by ocean currents has fascinated many famous explorers, including Charles Darwin and Thor Heyerdahl. Geodetic Survey using stoppered bottles containing a numbered postcard. When a bottle is found on a beach the finder fills out the card and drops it in the mail. It takes about one year for a drift bottle to float from Yucatan to Ireland. A bottle launched near Caracas, Venezuela reached the Florida Keys four months later, traveling at an average speed of 16 statute miles per day. It is estimated that tropical seeds found on European shores probably have been adrift for a year or longer.

During his famous voyage around the world on the H.M.S. Beagle, Charles Darwin championed the idea of drift seeds and fruits colonizing distant islands, particularly isolated volcanic islands that have never been connected to the mainland. Darwin studied the role ocean currents played in the flora of Cocos Keeling Islands in the Indian Ocean, and concluded that most of the endemic vascular flora was derived from drift seeds and fruits.

After he returned to England, Darwin conducted flotation experiments with cultivated plants. In the Journal of the Proceedings of the Linnaean Society Darwin stated: "I soon became aware that most seeds, in accordance with the common experience of gardeners, sink in water; at least I have found this to be the case, after a few days, with the 51 kinds of seeds which I have myself tried; so that such seeds could not possibly be transported by sea-currents beyond a very short distance." Darwin also mentioned rafting as a dispersal mechanism for seeds that generally don't float well in sea water. In addition, he stated that seeds contained within pods, capsules and the heads of Asteraceae may be carried by ocean currents and washed ashore on distant beaches. In his Origin of Species, 1859, Darwin summarized his experimental data on seed dispersal in salt water, and expressed a higher confidence in dried seeds: "Therefore it would perhaps be safer to assume that the seeds of about 10/100 plants of a flora, after having been dried, could be floated across a space of sea 900 miles in width, and would then germinate."

Of all the 250,000 species of seed plants on earth, only about 250 species (0.1 percent) are commonly collected as drift disseminules on tropical beaches; and only about half of these are known to produce seeds that can float in seawater for more than a month and still be viable. This relatively small number of drift seed species does not include seed plants which are dispersed on vegetation rafts, drift garbage from ships, or true marine seagrasses which live totally submersed in seawater. Although the total number of drift seed species with long viability periods may be relatively small, they nonetheless form a floral flotilla comprising countless thousands of individuals riding the ocean currents of the world.

The Hawaiian archipelago has been isolated from continental land masses during the past 30 million years, and yet the 1,000 species of indigenous Hawaiian angiosperms are believed to stem from natural introduction by long-distance dispersal of 280 ancestral plant about 14 percent of the original flowering plant immigrants to the Hawaiian Islands are clearly adapted to oceanic drift. If dispersal by birds and air currents are ruled out, it appears that seeds were carried thousands of miles to these islands, possibly by rafting or within protective capsules and pods. For example, California tarweeds are not included with tropical drift seeds, and yet an ancestral tarweed traveled at least 3,000 miles to the Hawaiian Islands where it gave rise to a remarkable group of endemics known as the "Silver Sword Alliance." The small seeds from ancestral members of the lobelia family (Campanulaceae) also reached these islands giving rise to an unusual group of Hawaiian lobelioids. One of these is the pachycaul Brighamia insignis that grows on steep sea cliffs on the island of Kauai.

An interesting coral tree endemic to dry, leeward slopes of all the main Hawaiian islands is called wiliwili (E. sandwicensis). This Hawaiian endemic belongs to a group of closely related passerine species in Erythraster, including E. tahitensis, endemic to Tahiti; E. variegata, a widespread species in tropical Asia and the South Pacific; and E. velutina, another widespread species in northern South America, the West Indies and the Galapagos Islands. Although the seeds of wiliwili sink in water, it may have originated from an ancestral species with buoyant seeds that reached these islands many thousands of years ago. In fact, a related tropical Asian species (E. variegata), which is commonly planted as a street tree in Hawaii, has seeds that float in water. The seeds of this species sometimes show up along Hawaiian beaches and tidal waters; however, they are usually of local origin. Trees of E. variegata are fairly easy to identify because the pods are longer than most cultivated coral trees, often up to 12 inches or more in length.

- Although the reasons are not identical, the loss of flight in certain insects and birds. "If a plant shifts its ecological preference, it will tend to lose contact with the agent responsible for its dispersal."

Seeds of the Asian coral tree (Erythrina variegata) are buoyant in seawater and may have drifted to distant shores of the tropical Pacific. In fact, this species (or its progenitor) may have given rise to several species of endemic coral trees in the tropical Pacific region.

THE HAWAIIAN SILVER SWORD

The Hawaiian islands are thousands of miles from the nearest continent or island mass. Even though they include some of the most isolated islands on earth, they contain numerous endemic species of plants, many of which evolved from ancestral seeds that reached the islands by drifting. Some striking members of the sunflower family evolved on the Hawaiian Islands from an ancestral California tarweed (Hemizonia) that colonized these isolated Pacific islands millions of years ago. This group of plants, called the Silver Sword Alliance, includes three genera and about 30 endemic species, an excellent example of adaptive radiation. One of the most amazing of all these Hawaiian endemics is the silver sword (Argyroxiphium sandwicense ssp. macrocephalum) that only grows in the cinders of Haleakala Crater on the island of Maui. The rosettes of sword-shaped leaves are covered with silvery hairs that reflect light and heat and provide insulation against the intense solar radiation and extreme aridity of this 10,000 foot (3,000 m) volcanic mountain.

A silver sword (Argyroxiphium sandwicense ssp. macrocephalum) in full bloom inside the lunar-like crater of Haleakala on the island of Maui. This remarkable plant and its relatives are descendents of an ancestral California tarweed (Hemizonia) that reached these islands millions of years ago, presumably by drifting.

NICKERNUTS

Some of the most ubiquitous drift seeds that readily germinate on tropical beaches are the gray and yellow nickernuts, Caesalpinia bonduc, C. ciliata, C. major. All are climbing, sprawling beach shrubs armed with vicious, recurved thorns and prickly pods bearing gray or yellow seeds. The smooth, marble-like seeds are commonly strung into necklaces and bracelets, often mixed with other colourful seeds. In Guayaquil, Ecuador, drilled nickernuts are sold by street vendors as amulets to ward off evil spirits. On some Caribbean islands nickernuts are used in a strategy board game called "Island Waurie." Two players sit on opposite sides of a hardwood board with six depressions (wells) on each side, each well with four nuts. Through a series of carefully planned moves, players attempt to empty all six of their wells before their opponent can do so. Legend has it that Island Waurie was originally introduced to the Cayman Islands by Blackbeard the Pirate on one of his voyages from South Africa. The game became popular with Cayman Islanders and was enjoyed by Ernest Hemingway whenever he visited the islands. Know in Africa as "mancala," the games were played for thousands of years in Egypt, where boards have been found carved into the stone of the pyramid of Cheops and the temples at Luxor and Karnak.

THE COCONUT

Probably the best known of all plant drifters is the coconut (Cocos nucifera). In fact, it is hard to imagine a tropical beach without coconut palms. The origin of the coconut it has never been found truly wild, every coconut palm is planted by man or derived from such a planting." "There is no island or shore where its presence is not due directly or indirectly to its having been planted by man." Probably under most situations, coconut seedlings require watering and other attention from humans; cite several localities where coconuts appear to have seeded themselves naturally, including cays in British Honduras (Belize), rocky islets in the Fiji group, the east coast of Trinidad, Cocos-Keeling Atoll in the Indian Ocean, and Krakatau and adjacent islets following the catastrophic eruption of 1883.

Coconuts have naturally established themselves on beaches of the tropical Pacific. but they all belong to either of two major types known as niu kafa and niu vai. The niu kafa types have an elongate, angular fruit, up to 6 inches in diameter, with a small egg-shaped nut surrounded by an unusually thick husk. Niu vai types have a larger more spherical fruit, up to 10 inches in diameter, with a large, spherical nut inside a thin husk. The niu kafa type represents the ancestral, naturally-evolved, wild-type coconut, disseminated by floating. The niu vai type was derived by domestic selection for increased endosperm ("meat" and "milk") and is widely dispersed and cultivated by humans. Both types of fruit can float, but the thicker, angular husk adapts the niu kafa type particularly well to remote atoll conditions where it can be found today. The presence of "undesirable" wild-type coconuts growing in mangrove swamps is clear evidence that they were self-sown and not planted by farmers. Angular, small-seeded coconuts on Ambergris Caye, off the coast of Belize. These coconuts resemble the wild-type niu kafa They were probably introduced to the Caribbean region of Central America by Portuguese traders presence of coconut palms on Cocos Island off the coast of Costa Rica and parts of the Pacific Coast at the time of Columbus. One of the proponents of the New World origin is Thor Heyerdahl, Heyerdahl expresses great confidence in Polynesian sailors who crossed vast stretches of the Pacific Ocean carrying staple foods from the New World, such as sweet potato, plantain and the coconut.

The sweet potato (Ipomoea batatas). Archaeological evidence shows that sweet potatoes were cultivated in South America by 2400 B.C. and fossilized sweet potatoes from the Andes have been dated at 8,000 to 10,000 years old. Although the sweet potato is clearly native to South America, it was cultivated in Polynesia as early as 1200 A.D. In fact, the sweet potato had already become the principle food in New Zealand by the time of Captain Cook's historic voyage to that part of the world in 1769. It is interesting to note that the sweet potato is known as "kumar" or "kumal" in the Lima region of coastal Peru, and it is called "kumaraHeyerdahl (1968) postulated that sweet potatoes were carried across the Pacific by Peruvian Indians before Europeans began to sail the world's oceans.

The coconut is indigenous to the Indo-Malaysian region. It spread by sea currents to many Pacific Island groups with adequate rainfall and moderate temperatures where the seedlings became established along beaches. Coconuts prefer the well-drained coral sand beaches of tropical islands and atolls, and are poorly established along shores of continents. Naturally dispersed coconuts can withstand occasional brief salt water flooding. Developing coconut palms obtain fresh water and mineral nutrients from a lens (soil layer) of fresh water (derived from island rainfall) which literally floats above the denser salt water beneath the beach sand.

3,000 miles seems to be the average maximum distance that a coconut will remain afloat and still remain viable. These limitations greatly diminish the chance of a coconut reaching the New World, let alone sprouting on a continental shoreproviding thatching, food, drink, coir, copra, oil, and hard endocarps which are fashioned into all sorts of decorative and useful articles. The origin of the generic name for coconut "Cocos" may be traced to the three germination pores on the endocarp layer surrounding the seed. Portuguese and Spanish traders introduced the coconut into West Africa after 1500. They called it "coco" from the Portuguese or Spanish slang word for monkey face, supposedly because of the eye pattern on the endocarp and the brown, fibrous hair (husk). Coconuts were later introduced into the Americas by these early traders. The center of origin for ocean-dispersed coconuts appears to be the Indo-Malaysian region.

Perhaps the best physical evidence for an Indo-Pacific origin of the coconut comes from New Zealand and India, where fossil specimens of a species of Cocos have been fossil coconuts of an extinct Miocene palm (Cocos zeylandica) have been found in North Island, New Zealand. Fossil evidence from such far-flung areas the coconut had ample time to disperse naturally in the Indo-Pacific, before humans came along to speed up the process.

THE GALAPAGOS ISLANDS

The Galapagos Islands, always a provocative place for students of natural history, offer a wide variety of examples of drift dissemination, including the independent transport of seeds and fruits floating in the ocean and the transport of seeds and animals on vegetation rafts from the mainland. Several species of mangroves grow on these islands, including the ubiquitous red mangrove (Rhizophora mangle) with its conspicuous prop or stilt roots descending from its limbs. The cigar-shaped disseminule, known as a "sea pencil," is unusual because it is a germinated seedling and not a seed or fruit. The seedling, with its long, pendant taproot (mostly hypocotyl), drops from the parent plant, where it may be automatically planted in the mud of shallow water or washed away to a distant shore. Beach vines, such as beach morning glory (Ipomoea pes-caprae), gray nickernut (Caesalpinia bonduc) and beach bean (Canavalia maritima), occur on several of the islands. Their durable seeds ride the currents of the world's oceans, resulting in enormous pantropic distributions.

Drift seeds and fruits from the shores of Floreana Island in the Galapagos Archipelago. A. Nickernut (Caesalpinia bonduc), B. Manchineel (Hippomane mancinella), C. Beach Morning Glory (Ipomoea pes-caprae), D. Sea Purse (Dioclea sp.), E. Sea Bean (Mucuna sp.) and F. Prickly Palm (Acrocomia sp.). D, E, and F are mainland species that probably drifted to this island.

It would take about two weeks for a floating log or vegetation raft to reach the Galapagos Islands from mainland Ecuador. Considering that the archipelago is over 3 million years old, the drift seed scenario offers a plausible explanation for how some plants made the 600-mile (965-km) journey.

Successful dissemination by drifting would not have to occur often for it to have a profound effect on the ecology of the islands. Taking into account all methods of dispersal, including transport by wind, birds, and drifting on the ocean surface, Duncan Porter, a botanist at Virginia Polytechnic Institute and State University, estimates that 378 original introductions could account for the 522 indigenous plant species known to grow on the islands. Of these natural introductions, 59% were a result of transport by birds, 32% by wind transport, and 9% by drifting on the ocean surface. If only one species successfully drifted to the islands and germinated there every 50,000 years, this would still account for the introduction of roughly 10 percent of the islands's flora., it would require only one species to arrive and become established every eight to ten thousand years.

Ancestors of present-day Galapagos reptiles, such as the remarkable marine iguanas, may have also drifted to the islands. In the swift current of Ecuador's Guayas River, trees, branches, and large mats of vegetation broken loose from riverbeds are commonly carried out to sea. Reptiles now on the Galapagos may have ridden such rafts from the mainland many thousands of years ago. A marine iguana (Amblyrhynchus cristatus) on the rocky shoreline of Hood Island in the Galapagos Archipelago. These unusual sea-dwelling iguanas are descendants of an ancestral mainland species that rafted to these islands thousands of years ago. In the distance is the infamous M/N Bucanero.

THE PANTROPICAL MANGROVES

Mangroves include several dozen species of trees and shrubs that grow along tidal swamplands throughout tropical regions of the world. Since their roots are emersed in water-logged silt and mud, often deficient in oxygen, the buttressed trunks and aerial roots of mangroves are typically dotted with numerous lenticels—small pores which provide gas exchange between the roots and the atmosphere. Lenticels are especially conspicuous on the prop roots of the ubiquitous red mangrove (Rhizophora mangle) and on the deeply-fluted trunk of the tea mangrove (Pelliciera rhizophorae), an interesting swamp tree of the tea family (Theaceae) along the Pacific coast of Costa Rica. The lateral roots of some mangroves produce upright, slender outgrowths called pneumatophores which extend above the mud and water like snorkel tubes. The porous pneumatophores are covered with lenticels and provide additional aeration for the root systems. The black mangrove (Avicennia germinans), of Central America and the Caribbean region, produces literally hundreds of pencil-like pneumatophores around its base.

Mangroves survive in seawater with a salinity that would be lethal to most trees and shrubs. Like celery or carrot sticks placed in saltwater, the roots of most plants rapidly lose water if they are suddenly emersed in seawater. Halophytes (salt-loving plants), such as mangroves, generally have a lower concentration of water molecules (lower water potential) in their root cells so they can take in water. They maintain lower water potentials in their roots by having higher internal salt concentrations than seawater and by losing water at the leaf surface. Since high internal salt concentrations can be lethal to plant cells, some species such as the black mangrove and white mangrove (Laguncularia racemosa), can excrete excess salt through special glands in their leaf blades and petioles. Red mangroves have root cell membranes which prevent the absorption of excess salt.

The seeds of mangroves are especially remarkable because they commonly germinate within their fruit while still attached to the parent plant, a condition known as "viviparous seeds." Having their embryonic root (hypocotyl) already elongated gives them a better chance of establishing themselves in soft mud during low tide. Called "sea pencils," the cigar-shaped seedlings (disseminules) of red mangrove may have an elongate taproot up to 10 inches long when they drop from the parent tree. The unusual, top-shaped fruit of tea mangrove contains one of the largest seeds in the world (excluding palms). It floats with the elongate, embryonic root pointing downward, and readily becomes implanted in soft mud. Germinated seedlings and sprouted fruits of the black mangrove and white mangrove are also dispersed by ocean currents. Red mangrove (Rhizophora mangle) with two viviparous (germinated) seedlings which have emerged from their fruits. The elongate embryonic axis called the hypocotyl develops into an extended taproot; Center: Onion-sized fruits of the tea mangrove (Pelliciera rhizophorae). The pointed beak encloses the embryonic root (hypocotyl). The fruit floats with the beak facing downward. When it becomes stranded in shallow water at low tide, the beak becomes implanted in the soft mud or silt, thus fascilitating the establishment of the seedling; Right: A sprouted fruit of black mangrove (Avicennia germinans) showing an elongated root.

Red mangrove (Rhizophora mangle) with a viviparous (germinated) seedling still attached to the fruit on the parent shrub. The seedling drops into the water where it is automatically planted in the soft mud or floats away.

POLYNESIAN BOX FRUIT

Beaches of French Polynesia are often littered with a buoyant drift fruit resembling a small coconut with flattened sides. It is called box fruit (Barringtonia asiatica) and is one of the most durable and widespread of all drifters, remaining buoyant for at least two years. In fact, they are used as fishing floats in Southeast Asia. Box fruits (Barringtonia asiatica) are widespread drift fruits in the tropical Pacific, remaining buoyant for more than two years. They are common in the turquoise-blue waters of French Polynesia

POLYNESIAN TAMANU

An equally common drift fruit of the tropical Pacific is called "tamanu" (Calophyllum inophyllum) by the Polynesians. The smooth, gray fruits resemble ping-pong balls strewn along wave-swept beaches. This tree belongs to the garcinia family (Clusiaceae syn. Guttiferae). " common drifter, called tropical almond (Terminalia catappa), resembles an oversized unshelled almond. It is one of the most distinctive trees of tropical beaches with tiered, pagodalike limbs and colourful red and yellow leaves reminiscent of a deciduous forest in autumn. Seeds of the Polynesian "tianina" or lantern tree (Hernandia nymphaeifolia) are produced in fleshy red or white "lanterns" which float like colourful boats in the clear blue water. The woody, seed-bearing endocarps are polished by native islanders and made into shiny brown leis and necklaces.

The smooth, ping-pong ball fruits of tamanu (Calophyllum inophyllum) commonly drift ashore on beaches of French Polynesia. On the atoll of Tetiaroa they are strung on fishing line to make unusual lamp shades. Flowers of tamanu (Calophyllum inophyllum). The stamens occur in five bundles. On the small East-Indonesian island of Alor, this pantropic tree is called "camplung." Here the sticky seed within the dried pericarp is used for lighting. It burns with a clean flame and reportedly fends off mosquitoes.

THE POLYNESIAN PANDANUS

Another widespread Polynesian plant called screw pine (Pandanus tectorius) is also dispersed by ocean currents. The large multiple fruit (resembling a pineapple) is composed of buoyant, one-seeded sections called "keys." The hard, woody keys are also polished and made into necklaces and leis.

The multiple fruit of Pandanus tectorius, showing the individual one-seeded sections called "keys." In addition to the edible seeds (one inside each key), the keys are polished and used for necklaces and leis. The keys are very buoyant and water-resistant, and remain viable for months. They are dispersed by ocean currents to shores of distant atolls and islands throughout the tropical Pacific.

Sea Beans

Some of the most beautiful of all drift seeds are called "sea beans." They come from large climbing vines (called lianas) throughout rain forests of the New and Old World tropics. The seeds are produced in large pods and resemble round, flattened beans with hard, woody seed coats. In species of Mucuna the pods are covered with whiskerlike, stinging hairs—presumably to discourage ravenous seed predators. The pods develop from clusters of greenish, bat-pollinated blossoms at the end of long, ropelike branches which hang from the forest canopy. Sea beans are also called "hamburger seeds" because they have two rounded halves enclosing a conspicuous central connection layer (hilum). In Mexico and Central America they are also called "ojo de buey" because of their uncanny resemblance to the eye of a steer. Another species of sea bean (M. argyrophylla) called "ojo de venado" (deer eyes) has a most remarkable use in Central America. he sex of the seed is determined by whether they sink or float in water. Those that sink in water are called "hembras" (female) and those that float are "macho" (male). The hemorrhoidal treatment requires the sufferer to carry a "male" and a "female" seed in their back pocket. Some species of sea beans are called sea purses (Dioclea) because they are shaped like a purse, including a circular hilum along the edge that superficially resembles a zipper.

Velvety pods of the sea bean (Mucuna argyrophylla) hang from long, ropelike stems in the rain forests of Belize. The hard, black seeds are called "ojo de buey" (eye of the steer) and "ojo de venado" (eye of the deer) by local residents.

The Sea Heart

Perhaps the most remarkable of all drift seeds resemble large wooden hearts and are called "sea hearts" (Entada gigas). The heart-shaped seeds are produced in huge bean pods up to six feet long, the longest of any legume. In the rain-soaked tropical forest near Golfito, Costa Rica, enormous sea heart lianas twine through the forest canopy like a gigantic botanical boa constrictor. The vine is known locally as "escalera de mono" or monkey ladder because it provides a maze of arboreal thoroughfares for New World monkeys. Torrential rains wash the seeds into streams and rivers where they reach the sea, and perhaps eventually the shores of a distant continent. Sea hearts are highly prized by beach combers and make beautiful pendants when polished. Sea hearts and a similar rectangular Old World species (E. phaseoloides) were commonly used in Norway and northern Europe for snuffboxes and lockets. The seeds were cut in half, the contents removed, and the woody seed coats hinged together.

Sea hearts have a long and colourful history in fact and fiction. Early naturalists thought the unusual seeds came from strange underwater plants whose origin was shrouded in mystery. In England sea hearts were carried as good luck charms by sailors embarking on a long ocean voyage. It was thought that if sea hearts could survive a long and perilous journey across the ocean, perhaps they could also protect their owner. Christopher Columbus was fascinated with objects that drifted ashore on beaches of the Azores, off the coast of Portugal. It is said that a sea heart provided inspiration to Christopher Columbus and led him to set forth in search of new lands to the west. In fact, the sea heart is called "fava de Colom," or Columbus bean, by Portuguese residents of the Azores.

The Mary's Bean

Another drift seed called Mary's bean (Merremia discoidesperma) was a special find to pious beachcombers. Named after the Virgin Mary, it is also called crucifixion bean because of a cross etched on one side. The distinctive seeds are produced in papery capsules on a climbing tropical vine. The cross is actually an impression where the seed was attached inside the capsule. Because of a thick, woody seed coat and internal air cavities, the seeds are very buoyant and may drift for months or even years at sea. Historically, the unique seeds have been used for good luck charms and to ward off evil spirits. A woman in labor was assured an easy delivery if she clenched a Mary's bean in her hand. Seeds were handed down from mother to daughter as treasured keepsakes. To this day Mary's beans are occasionally sold by street vendors in Costa Rica.

Some drift seeds and fruits on the Caribbean shores of Costa Rica. A. Starnut Palm (Astrocaryum sp.), B. Mary's Bean (Merremia discoidesperma), C. Oxyrhynchus trinervias, D. Crabwood (Carapa guianensis), E. Prioria copaifera, F. Sea Bean (Mucuna sp.), G. Sea Coconut (Manicaria saccifera), and H. Calatola costaricensis.

In its native habitat, the Mary's bean is known from only a few locations in Mexico and Central America. It is a lovely vine of the Morning Glory Family (Convolvulaceae) with beautiful yellow, funnel-shaped blossoms. As drift seeds, the Mary's bean is known from the Marshall Islands to beaches of Norway, a total distance of more than 15,000 miles (24,000 km). this constitutes the widest drift range of any seed or fruit which has been documented. Other pantropical drift seeds may drift as far or farther, but their precise point of origin cannot be determined with certainty.

Large Neotropical Drift Fruits

With the exception of introduced cattle, donkeys and horses, no native mammals of the New World tropics can crush the hard, thick-walled pods of many rain forest trees in their jaws. Livestock apparently like the sweet pulp inside the pods of West Indian locust (Hymenaea courbaril) and disperse the hard, viable seeds in their excrement. In areas without livestock, the rotting pods litter the ground beneath large trees. Agoutis, tapirs and peccaries chew open the rotting pods and eat the sweet pulp and seeds, but are not major agents of seed dispersal like the larger hoofed mammals. large grazing mammals, including extinct pleistocene elephants called gomphotheres, may have once eaten the pods and dispersed the seeds in lowland forests. In Africa, the large woody pods of related species are quickly devoured by large herbivores. There are other Central American rain forest trees that also appear to be missing their natural herbivorous dispersal agents. Their hard, woody, indehiscent fruits pile up beneath the branches and slowly rot away in the soggy, moldy layer of soil and debris. The lack of natural dispersal agents may also apply to devil's claw pods that litter the ground in temperate regions of the New World.

Hard, woody fruits of some Central and South American rain forest trees were likely dispersed by large prehistoric herbivores thousands of years ago. A. Cassia grandis (legume family—Fabaceae); B. Crescentia alata (bignonia family—Bignoniaceae); C. Hymenaea courbaril (legume family—Fabaceae); and the palms (Arecaceae): D. Attalea speciosa (one fruit cut open to reveal thick, woody pericarp); E. Raphia taedigera; and F. Orbignya cohune.

The Manchineel Tree

There are dozens of unusual drift seeds that wash ashore on Caribbean islands, each with fascinating stories about them. One of the most interesting is an intricately-sculptured star that comes from the fruit of a native tree called the manchineel tree (Hippomane mancinella). The fruit of the manchineel tree resembles a small green apple, except this "apple" has a woody, seed bearing endocarp and is quite poisonous. Manchineel trees are common along Caribbean shores, and the fruits often fall into the water. The outer fleshy layer rots away and the buoyant, woody endocarp starts a new career as a drift disseminule. After many months of wave action and abrasion in tropical waters, the endocarp gradually becomes eroded into an intricate star. It is a beachcomber's delight to find one of these beautifully-sculptured stars. Fruits of the manchineel tree (Hippomane mancinella) resemble small green apples. The seed-bearing pits (endocarps) float in tropical waters and gradually become eroded into intricately-sculptured stars.

The manchineel tree belongs to the diverse Euphorbia Family (Euphorbiaceae), and like many members of this family, it contains a poisonous milky-latex sap. The toxin is a mixture of diterpene esters, and contact with the skin may cause inflammation and a blistering rash. The caustic terpenes are structurally similar to phorbol, a potent carcinogen listed among organic There were wild fruits of various kinds, some of which our men, not very prudently, tasted; and upon only touching them with their tongues, their countenances became inflamed, and such great heat and pain followed, that they seemed to be mad, and were obliged to resort to refrigerants to cure themselves."

A young manchineel tree (Hippomane mancinella) that has sprouting in the sand of a Costa Rican beach. The beach was littered with the green, apple-like fruits of this ocean-dispersed species.

SEA COCONUTS AND HAND GRENADE POD

Two more interesting drift fruits that wash ashore along beaches of the Caribbean and the southeastern United States are sea coconut (Manicaria saccifera) and grenade pod (Sacoglottis amazonica). Both drift fruits come from trees that are native to the Amazon River, and are often carried by the Gulf Stream to beaches of Northern Europe. The sea coconut is a tall, unusual palm with leaves nearly 30 feet (8 meters) long. the leaf is typically divided shallowly at the tip and is not truly entire. The tuberculate fruit wall contains one to three globose seeds which become sun-bleached and resemble golf balls strewn along beaches. This striking palm is more closely related to the coco-de-mer (Lodoicea maldivica) of the Seychelles Islands rather than the true coconut (Cocos nucifera).

A three-seeded fruit and two seed-bearing endocarps of the sea coconut (Manicaria saccifera), a palm native to the Amazon basin. Sometimes the tuberculate fruits contain only one or two seeds. The buoyant fruits of grenade pod contain numerous hollow pockets which permit them to float for great distances. The raised pocket areas, which are exposed when the outer exocarp layer is eroded away, produce the faceted (bumpy) surface of the endocarp which superficially resembles a hand grenade. In Mexico these unusual fruits are often called "cojon de burro," a descriptive name that is probably not as appropriate as hand grenade. They are produced by a small tree native to the Amazon and Orinoco River basins of South America.

Like the migrations of ancient mariners, seafaring seeds have traveled to tropical beaches throughout the world. Collecting drift seeds can be a rewarding hobby—botanical treasures from exotic beaches to be admired for years to come. Some seeds (such as sea beans and sea hearts) can be polished in a stone tumbler, as you would polish agates and other precious stones, and different seeds can be strung together to make attractive necklaces and leis. Who knows—they might even bring you health and good fortune, like the fabulous tales from courageous explorers many centuries ago.

DRIFT SEEDS AND DRIFT FRUITS

Seeds provide the vital genetic link and primary dispersal agent between successive generations of plants. They are produced and packaged in botanical structures called fruits, and come in an endless variety of shapes and sizes. Tropical drift seeds and fruits are especially remarkable because they can survive months or even years at sea. They are very buoyant, with thick protective shells which are impervious to salt water. In some drift fruits, such as the coconut, the seed embryo and fleshy white "meat" (endosperm) is enclosed within a hard, bony layer (endocarp) surrounded by a thick fibrous husk. Other drift seeds have thick woody seed coats and internal air cavities which produce their buoyancy. During their long voyages they often cross entire oceans, perhaps colonizing the shores of a coral atoll or isolated volcanic island. Sea dispersal is a hit-or-miss method; after a long, perilous journey, other dangers await the vulnerable seedlings. They may perish in the parched sand without adequate moisture or be quickly devoured by ravenous land crabs. Drift seeds carried by the Gulf Stream from the Caribbean to beaches of northern Europe may find themselves in another precarious situation—aliens in a hostile climate much too cold for them to survive.

The tropical islands of Indonesia, Polynesia and the Caribbean probably have the greatest variety of drift seeds. They are especially abundant along wave-swept beaches following the hurricane season, and may be carried inland by tidal waters. The Cayman Islands and Yucatan Peninsula lie directly in the path of the Equatorial Current, bringing thousands of unique drift disseminules from the Amazon and Orinoco deltas and across the Atlantic from river deltas of tropical west Africa. In the Pacific region drift seeds may be deposited well above the inhospitable beach zone by enormous tidal waves known as tsunamis. The tsunami emanates from an earthquake or volcano epicenter and moves at great speed across the Pacific Ocean. Drunk Bay on the windward side of the island of St. John in the U.S. Virgin Islands. The following variety of drift seeds and fruits came from Drunk Bay.

THE LONGEST DISTANCE TRAVELED BY A DRIFT SEED

Imagine floating helplessly in the open sea, thousands of miles from land, your destination at the mercy of the wind and currents. Eventually. This scenario is precisely what happens to countless drift seeds and fruits, a remarkable flotilla of flowering plants that travel the oceans of the world. The world's record for the longest distance traveled is difficult to determine. Some widespread drift seeds, such as sea beans (Mucuna) and sea hearts (Entada) have probably floated longer distances in the sea; however, most of these drift seeds have pantropical distributions and their precise point of origin cannot be determined. In addition to long ocean voyages, the sea heart (Entada gigas) also produced the longest bean pod of any member of the legume family (Fabaceae).

But there is one drift seed with a very limited point of origin called the "Mary's bean" (Merremia discoidesperma). Named after the Virgin Mary, it is also called the crucifixion bean because of a distinctive cross etched on one side. The unique seeds are produced by a tropical liana of the morning-glory family (Convolvulaceae) that is only known from relatively few locations in the rain forests of southern Mexico and Central America. As an ocean drift seed, the Mary's been is known from Wotho Atoll in the Marshall Islands to the beaches of Norway, a total distance of more than 15,000 miles. The Mary's bean (Merremia discoidesperma) is certainly one of the most elusive and interesting of all drift seeds in fact and fiction. A thick, woody seed coat and internal air cavities enable this remarkable seed to drift for years at sea, from Central America to beaches of Norway.

POLLINATION AND FERTILISATION

Seeds are the result of fertilisation of the ovules by pollen. The ovules are 'half a seed' - they contain the DNA of the female parent plant (sometimes called 'the seed parent'), but cannot develop into complete seeds until they have had the DNA of the male parent plant (sometimes called 'the pollen parent') added.

The DNA is contained on chromosomes, and when the ovules or stamens are being made, the plant's DNA divides in half, and each parent puts a set of its own chromosomes into its half of the seed (the ovules for the female parent, and the pollen for the male parent). When the ovules are fertilised, the two sets of chromosomes join together to make the double set which a new plant needs. The fact that the same genes are passed on to each succeeding generation is what makes the new plant similar to its parents, so daisies produce daisies in the next generation, and do not have the DNA to make a tree, for instance. For fertilisation to take place, pollen of the right type must join up with the ovules. The pollen is contained in bags called pollen sacs which open when the pollen is ripe. Some pollen is only ripe for a few days, some lasts much longer. We think of pollen as yellow powder, but it can be white, orange, brown, purple or black, and it comes in many different shapes and sizes. Botanists are able to identify a plant by the shape of the individual pollen grains! In most plants, each flower contains both an ovary and stamens. The plant contains both male and female parts. This is the centre of an individual flower on a Honeysuckle. There are five yellow stamens, and a round green stigma leading to an inferior ovary which will develop into a berry. The ovary can be either inside the flower (called superior) or behind the flower (called inferior), and can contain just one or many ovules.

the stamens are on one flower, and the ovules are in another flower. Sometimes, a plant produces only male flowers, that is, flowers which have only stamens, so they will not produce seeds - the Holly is an example. The female flowers, those with a stigma leading to ovules, are on a separate plant. In the Holly, a separate male plant is needed to pollinate the flowers so the female plant can produce berries. Plants with both male and female parts in the same flower, or with male and female flowers on the same plant, are called 'monoecious'. Plants that have either all male or all female flowers are called 'dioecious'.

These flowers are Malva sylvestris - the one on the right is the male flower - you can see the white stamens clustered together. The flower on the left is the female flower - it has a very divided, curly purple stigma. Here they are in close-up:When pollen with the right chemical password lands on the stigma, it grows down the style to reach the ovules. In the picture, the yellow tube shows the journey the pollen grain has to make to reach the ovary. The ovules are shown as pale yellow circles - the real ovules are too small to be seen at this stage. The ovules can then grow on to produce the seeds.

The stigma is only able to receive the pollen for a limited time. When it is ready, the tip is sticky to help the pollen grains stay in place. Stigmas also come in many shapes, sizes and colours. The stigma is on the end of a stalk called the style. Sometimes the style is long and thin, and the stigma on top can be rounded like a pin. In Primroses, if you can see the round stigma in the throat of the flower, it is called 'pin-eyed'. When the short stamens are visible in the throat, it is called 'thrum-eyed'. Sometimes, the style and stigma are short and fat, as in a Poppy, or the stigma may be divided. The style and stigma can be green, or pink, or orange - in some cases, they match the colour of the flower. The pollen gets on to the stigma in many ways. Sometimes, it is blown from the stamens of one flower or one plant onto the stamens of another flower. Sometimes, the pollen is carried from flower to flower by insects. To attract pollinating insects, many flowers produce sweet nectar. These flowers often have long tubes, and the insects that pollinate them need long tongues. Larger flowers of this type are sometimes pollinated by humming birds, which also have long tongues. Some flowers are pollinated by moths, and these often have pale colours or perfumes which are stronger at night.

Pollinating insects get dusted with pollen as they brush against the stamens while trying to collect the nectar, and when they visit another flower to drink more nectar, the pollen is deposited on the stigma of the new flower. A flower that is pollinated by the pollen of another flower is cross-pollinated. When the pollen comes from the same flower, the flower is self-pollinated.Each seed is unique, formed from one ovule and one grain of pollen, with its own particular combination of all the separate genes from each of its parents. Because of this diversity and the possibilities it gives the next generations, nature sometimes takes precautions to prevent flowers being pollinated by their own pollen. Sometimes, the stigma won't accept pollen from the same flower or the same plant, sometimes the pollen ripens at a different time from the stigmas, or the stigmas are placed above the stamens so that it is more difficult for the flowers to be self-pollinated.

On the other hand, some plants have their stigmas and stamens close together, so that cross-pollination is easier, and some plants can produce seeds without the flowers opening at all. These are called cleistogamous seeds.

THE WORLD'S SMALLEST & LARGEST SEED

Certain epiphytic orchids of the tropical rain forest produce the world's smallest seeds, up to 35 million per ounce. One seed weighs about one 35 millionths of an ounce (1/35,000,000) or 0.81 micrograms. Some seeds are only about 1/300th of an inch long (85 micrometers). [The resolving power for an unaided human eye with 20-20 vision is just under 0.1 mm.] Orchid seeds are dispersed into the air like minute dust particles or single-celled spores, eventually coming to rest in the upper canopy of rain forest trees. The world's largest seed comes from the coco-de-mer palm (Lodoicea maldivica), native to the Seychelles Archipelago in the Indian Ocean. Although it belongs to a different genus from true coconut palms (Cocos), this enormous seed is often called the "double coconut." A single seed may be 12 inches (30 cm) long, nearly three feet (0.9 m) in circumference and weigh 40 pounds (18 kg). It should be noted here that the largest seed does not have the largest embryo. In fact, palm seeds are mostly composed of endosperm tissue and generally have relatively small embryo. Small seeds. Mustard family (Brassicaceae): Black mustard (Brassica nigra). Orchid family (Orchidaceae): Coralroot orchid

(Corallorhiza maculata). Duckweed family (Lemnaceae): Watermeal (Wolffia angusta), a one-seeded fruit called a utricle. Poppy family (Papaveraceae): Opium poppy (Papaver somniferum). Without any doubt, the orchids have the record for smallest seeds. The seeds of some species are no larger than fungal spores and occur in a loose cellular sheath. Since the seeds have no endosperm and underdeveloped embryos, there are practically no food reserves. In order to germinate under natural conditions, they must establish a symbiotic relationship with a compatible mycorrhizal fungus. During early stages of development, the fungus supplies critical nutrients to the orchid seedling. Later the orchid may become fully independent, or it may retain its mycorrizal relationship throughout its life. The above coral-root orchid seed (Corallorhiza) grows into a nonphotosynthetic mycotrophic wildflower. It absorbs carbohydrates and minerals from its fungal host, which in turn absorbs these vital nutrients from the roots of nearby forest trees. Orchid seeds are also grown under aseptic conditions in nutrient agar, similar to bacterial and fungal cultures. Wolffia certainly has the record for smallest fruits which are not much larger than grains of ordinary table salt (NaCl). The single seed inside is almost as large as the fruit; therefore, wolffia seeds are not as small as orchid seeds.

Microscopic view of the seed of a coral-root orcid (Corallorhiza maculata). The individual seed is only about 0.2 mm in diameter. In fact, there are unusual bacterial cells that are larger than this orchid seed. The resolving power for an unaided human eye with 20-20 vision is about 0.1 mm. With its cellular sheath (seed coat) removed, this seed is barely visible to the naked eye. Certain epiphytic orchids of the tropical rain forest produce the world's smallest seeds weighing only 35 millionths of an ounce.

One seed capsule from a single flower may contain up to four million seeds. They are dispersed into the air like minute dust particles or single-celled spores, eventually coming to rest in the upper canopy of rain forest trees. The seeds of some species are no larger than fungal spores and occur in a loose cellular sheath. Since the seeds have no endosperm and a minute, undifferentiated embryo, there are practically no food reserves. In order to germinate under natural conditions, they must establish a symbiotic relationship with a compatible mycorrhizal soil fungus. During early stages of development, the fungus supplies critical nutrients to the orchid seedling. Later the orchid may become fully independent, or it may retain its mycorrhizal relationship throughout its life. The above coral-root orchid seed (Corallorhiza) grows into a nonphotosynthetic mycotrophic wildflower that is completely dependent on its mycorrhizal fungus. Throughout its life, the orchid absorbs carbohydrates and minerals from its fungal partner, which in turn absorbs these vital nutrients from the roots of nearby forest trees. In a laboratory, orchid seeds can be grown in nutrient agar, like a sterile (axenic) culture of bacteria or fungal spores.

The seed pod (capsule) of an unknown orchid containing many thousands of minute seeds. Each seed is enclosed in a cellular sheath (seed coat) resembling a short, silky hair. The seeds are dispersed into the wind like dust particles. In nature, the probability of an orchid seed finding a suitable place for germination and a compatible fungal partner are unlikely, so millions of seeds are released to increase the the largest seed embryo goes to Mora oleifera (Fabaceae), a large tree that grows in tidal marshlands and estuaries along the Pacific coast of tropical America. In Costa Rica, this tree often forms nearly pure stands just behind the mangrove swamps. Seeds of M. oleifera may be up to 7 inches (18 cm) long and up to 5 inches (8 cm) wide. Another species (M. excelsa) has slightly smaller seeds. Like other exalbuminous legume seeds, the two cotyledons comprise most of the seed. Since the cotyledons are part of the embryo, this species is certainly a strong contender for the record of world's largest seed. The seeds float in ocean current with their two large cotyledons connected or separate. Dried cotyledons washed up on beaches superficially resemble the shells of a bivalve mollusk.

Seed of Mora oleifera showing the two large cotyledons. Since the cotyledons are technically part of the embryo, this is perhaps the largest embryo of any seed. Mora oleifera is a large tree that grows in tidal marshlands along the Pacific coast of tropical America. The seeds float in ocean water with their two large cotyledons connected or separate. Dried cotyledons that wash ashore on beaches superficially resemble the shells of a bivalve mollusk

Provocative seed and immature fruits of the Seychelles Island Palm (Lodoicea maldivica). Like the coconut (Cocos nucifera), the seed is enclosed by a thick, woody endocarp. The complete fruit (drupe) contains of an enormous, seed-bearing endocarp surrounded by a husk composed of a thin mesocarp and a smooth, outer exocarp. This is truly the largest seed produced by any plant on earth.

THE WORLD'S OLDEST GERMINATED SEED

Date palm (Phoenix dactylifera) that an archaeological excavation at King Herod's Palace on Mount Masada near the Dead Sea. Nicknamed "Methuselah" the palm that sprouted from this ancient seed has been Carbon-dated at about 2,000 years old. Since date palms are dioecious, it is not known whether the palm is male or female. Another seed viability record is the Asian water lotus (Nelumbo nucifera) in which a seed from China was successfully germinated after 1,200 years. But the world's record is the seed of an Arctic lupine (Lupinus arcticus) that was excavated from a lemming burrow in frozen Arctic tundra. The seed germinated and flowered after an estimated 10,000 years of dormancy.

THE WORLD'S LARGEST FLYING SEED

Flying through the air is another effective adaptation for fruit and seed dispersal by plants. Airborne seeds have several ingenious methods of flying through the air, including whirling like a helicopter, gliding, and floating like miniature parachutes with tufts of fine hairs. Of all the types of "helicopter seeds," those of Gyrocarpus are the most remarkable. The climbing gourd (Alsomitra macrocarpa), native to the Sunda Islands of the Malay Archipelago, produces one of the largest winged seeds up to 5 inches (13 cm) wide inside a large, club-shaped gourd. The football-sized gourds hang from a vine high in the forest canopy, each packed with hundreds of winged seeds.

Alsomitra is one of the most unusual members of the diverse gourd family (Cucurbitaceae). The seeds have two papery wing membranes and become airborne like a glider when released from the fruit. This large, streamlined seed reportedly inspired the wing design of some early aircraft, gliders and kites. Although the seeds vary in shape, some of the most symmetrical ones superficially resemble the shape of the "flying wing" aircraft or a modern Stealth Bomber. Large airborne seeds would not be complete without mentioning the quipo tree (Cavanillesia platanifolia), a massive rain forest tree in the bombax family (Bombacaeae) native to Panama. The enormous winged fruits of the quipo tree flutter through the air, carpeting the ground beneath the huge canopy of this striking tropical tree. Although it is large, this seed-bearing structure is the actual fruit, and not an individual seed as in Alsomitra.

The quipo tree (Cavanillesia platanifolia), a remarkable rain forest tree in the bombax family (Bombacaceae) with huge winged fruits. This massive tree is native to Panama.

7

Developments in the Seed Industry

Many major food crops upon which we depend originated and were first cultivated in Third World countries rather than in the industrialized countries where they are most productively exploited today and in which they have been improved by traditional plant breeding. The potato originated in South America, wheat in Ethiopia and the Near East, important maize varieties in South America, and rice in Asia. Moreover the so-called Vavilov centres, which contain the greatest variety of living species and which have been the source of much important gene material, are mainly in Third World countries.

The wealth of the industrial countries owes much to their past ability to exploit agriculturally crops and animals originating in other countries. In the colonial era the success of British scientists in smuggling rubber plants from Brazil to Sri Lanka, Singapore and Malaysia and the comparable more complex route by which coffee was introduced to Latin America from Ethiopia are instances of colonial powers obtaining significant wealth by exploiting plant material from the Third World.

In one way or another the process by which institutions and firms in developed countries have continued to collect new plant and animal varieties from less-developed countries has continued. But what is still a major source of friction is the procedures by which legislative protection is developed to give companies in the richer countries commercial rights to exclusive exploitation of varieties developed from plants freely collected from other countries, and to extract high rates of profit from the sale of those varieties. Most extreme is the situation whereby under the USA Plant Patent Act of 1930, which covered asexually propagated plants such as certain fruits, flowers, ornamental shrubs and trees, it has been possible for plants discovered and smuggled out of other countries to be given patent protection in the USA.

The key to this is that the Act accepted the unfeasibility of requiring proof of an 'inventive-step' for the granting of a patent, and employed the criterion that as far as can be determined the plant variety is new. Since some novel variety found in the rain forests of South America may satisfy this criterion, and the source of the plant can be concealed from the authorities, the theft and smuggling of plants can be said to have been encouraged. Seed companies in developed countries have successfully pressed their case for protection, and have progressively extended Plant Breeders' Rights (PBR) through both national legislation and the Union for the Protection of New Varieties of Plant (UPOV), which is an international agreement among signatory countries to honour a system of varietal rights. In the case of food crops, in the absence of patentability, PBR operates by registering distinct varieties. Pressure has been exerted upon less-developed countries to become signatories of UPOV and thus acknowledge the rights of breeders, most of whom are from developed countries. However, the majority have stalled or rejected membership, and have been fundamentally unhappy about accepting the principle of creating monopoly rights for others over plant material which in some cases originated in their own country.

The whole issue of intellectual property rights has been raised to a prominent political level by the USA's insistence that it be included as one of the thirteen areas for negotiation in the Uruguay Round of multilateral bargaining of the General Agreement on Trade and Tariffs. While the issue is a much broader one than that of plant and animal breeders' rights, it reflects a basic political conflict between the richer countries, determined to obtain secure returns on R & D expenditures by companies, and LDCs struggling to catch up and anxious to avoid having to pay royalties on products protected by PBR and patents. This conflict has broader moral and ethical dimensions when it relates to the issue of charging royalties for the seeds of food crops to countries where malnutrition is widespread.

Against the previous argument has to be balanced the need of those incurring R&D expenditure to capture enough of the returns to provide an incentive for R & D. At least that is certainly the case where R & D is to be undertaken by the private sector, which is politically the increasingly preferred solution in the UK, the USA and elsewhere. This, however, leads to another area of concern, which is that a small number of large multinational chemical and pharmaceutical firms have increasingly come to control the seed industry. Companies such as Shell, Sandoz, Dekallb-Pfizer, Ciba-Geigy and BP have become owners of most of what were small independent seed companies.

This linking between chemicals and seeds within firms has a strong commercial logic. It facilitates the marketing of packages consisting of seed varieties plus the appropri-ate fertilizers, insecticides, etc. It also allows the companies concerned to integrate their plant breeding, biotechnology and chemical research. One particular synergy for these companies will be to bioengineer varieties of major food crops to resist their own herbicides and weedkillers. This would permit chemical weed control to be extended to large-scale crops where this is not now possible, and is just one way in which the direction of biotechnolo-gical research may be influenced (biased) by the combination of forces giving a small number of companies extensive control over the basic means of agricultural production. There are many other issues which could be touched on regarding corporate influence over the development of agricultural biotechnology. It is, however, understandable that the spokesmen for LDC countries, which are profoundly concerned about their dependence upon western technology and exercised by what is seen as colonial and post-colonial exploitation by the West, should be alarmed by the prospect of new rights being created over seeds and other biotechnological products. Certainly there are widespread fears that they will only gain access to the fruits of biotechnology on unfavourable terms and that their relative dependence on and subservience to western industry will be increased so that benefits to them will be small. There are many western critics who would agree with this, and that the system by which agricultural biotechnology is delivered will favour the 'haves' rather than the 'have-nots'.

SEED DEVELOPMENT

One hundred fifty years ago the United States did not have a commercial seed industry; today we have the world's largest. Some view this as real progress, a form of genetic Manifest Destiny. A nation once a 'debtor' in plant genetics now supplies the world. In 1854, seeds were sourced in the U.S. by way of a small number of horticultural seed catalogs, farmer exchange, on-farm seed saving, and through the beneficence of the United States government. Specifically, beginning in the 1850s, the U.S. Patent and Trade Office and congressional representatives saw to the collection, propagation and distribution of varieties to their constituents throughout the states and territories. The programme grew quickly so that, by 1861, the PTO had annual distribution of more than 2.4 million packages of seed (containing five packets of different varieties). The flow of seed reached its highest volume in 1897 with more than 1.1 billion packets of seed distributed.

The government's objectives in funding such a massive movement of seed stemmed from the recognition that feeding an expanding continent would require a diversification of foods. To the early colonies, the introduction of wheat, rye, oats, peas, cabbage and many other vegetable crops was as critical to food security as was the adoption of the corn, beans and squash. Immigrants were encourage to bring seed from the old country, founding fathers such as Thomas Jefferson engaged in seed-exchange societies, and by 1819 the U.S. Treasury Department issued a directive to its overseas consultants and Navy officers to systematically collect plant materials.

The first commercial seed crop was not produced until 1866—cabbage seed produced on Long Island for the U.S. wholesale market. The industry flourished to some degree, but early seed trade professionals felt their growth was stymied by the U.S. government programs as well as the self-replicating nature of their product (that is, the factory contained within that product). In 1883, the American Seed Trade Association (ASTA) formed and immediately lobbied for the cessation of the government programs. The organization developed powerful allies, such as Grover Cleveland's Secretary of Agriculture, J. Sterling Morton, who wrote that the government giveaway was "antagonistic to seed as a commodity-form and in direct competition with the private seed trade." But the programme was very popular with constituents, and the USDA's seed budget was kept intact – at one point counting for a full 10 per cent of the agency's overall annual expenditures.

In the early part of the 20th century, the first wave of hybrids began to provide seed companies with a potential increase in product profitability (as farmers would now need to return to the seed distributor for materials each year). However, most of the hybrid development was occurring at Land Grant Universities, and these universities refused to give the companies exclusive rights to the seed. Once again, the industry felt its growth hindered by federal programs and complained of unfair trade practices. Mounting data also indicated a slowing in yield increases from seed developed in government programs. The industry used this last point to strengthen its argument for the privatization of seed development in order to foster greater food security.

In 1924, after more than 40 years of lobbying, ASTA succeeded in convincing Congress to cut the USDA seed distribution programs. The USDA still supported breeding at the state agricultural schools, and for a time these programs continued to compete with seed companies by developing 'finished' commercial varieties. Associations such as the American Society of Agronomy and American Society of Horticulture Science eventually convinced the public programs that their appropriate role was in training plant breeders, performing fundamental research, and creating raw materials and technologies for private industry to capitalize on. The LGUs began to increasingly serve in this capacity, developing inbred parental lines and breeding stock that the seed trade would use to create proprietary varieties.

These changes in the public role, along with improvements in hybrid techniques, led to the growth of the seed trade following World War II. The trade was well represented during this period by regional companies. The conversion to monocropping and large-scale corporate agriculture had not yet moved into full swing. The Santa Clara Valley grew vegetables and fruit and not internet startups, and Americans still planted their Victory Gardens. The seed trade reflected this diversity in food production.

In the 1960s, a few larger seed firms began to purchase smaller companies. But the consolidations of this period were minor compared to the frenzy that would come with a Supreme Court ruling on June 16, 1980, in the case of Diamond v. Chakrabarty. Prior to the Chakrabarty decision, a plant (or animal) could be owned, but the genetics could not. This case cleared the patenting of life forms on the bases of their genetic coding. The PTO granted more than 1,800 such patents following the ruling. Companies that had no historical seed interests—primarily chemical and pharmaceutical firms—began purchasing seed companies. In a few short years, there were billions of dollars in mergers and acquisitions—with little to no regulatory oversight—creating for the first time a majority ownership of plant genetics by a few multinational companies. No other natural resource (marine, timber, minerals) has ever shifted from public to private hands with such rapidity, such intensity of concentration, and so little oversight.

IMMEDIATE AND THE FUTURE

"There is a direct threat to our food system when we have a preponderance of genetic resources controlled by institutions whose only goal is profit," plant breeder Frank Morton expressed emphatically when asked for his perspective on the Monsanto acquisition. He went on to compare the present with the past, "When these services [breeding and production] were diffused amongst many individuals and groups with diverse motives, we had a much more diverse and healthy food system."

Diversity and competition have historically made for healthier economies as well as ecologies. Ecological and economic systems have another thing in common; as one group abandons a niche, it leaves room for others to move in and utilize it. Many of the seed company representatives and breeders I spoke with felt that the organic community can and should develop a less centralized seed system. Steve Peters, seed procurement manager for Seeds of Change, shared his firm's vision: "Part of our strategy is to go into neglected markets and respond to these needs. We want to offer true organic alternatives; this means we not only pay attention to the regional needs but also have a different approach to breeding – building new alliances in crop development. We're not chasing single-gene resistance but looking at horizontal resistance in traits like Downy mildew in spinach."

Adaptability has historically been an integral part of the organic movement– responding to customer inquiry and opinion, personal service (a face on the food), localized and decentralized – all traits that help make the organic sector healthy and promote its vigorous growth. "The organic seed world is where the organic produce market was twenty years ago," said veteran organic grower and seed producer Nash Huber.

"Produce quality wasn't always so great then, but we worked it out and now are better than the conventional systems. The potential in seed is that the customers – we farmers – will see that, in the long run, an organic seed trade will serve us." Farmers like Huber are investing in the organic seed trade. He works as an educator in the Organic Seed Alliance's WSARE organic seed production trainings, is leading a participatory plant breeding project with the Alliance, and produces seed for the organic seed trade as well as his own on-farm use. Carrots are the main cash crop on Huber's 400-acre farm, and he produces 80 per cent of his own seed to insure quality and availability. "I can't afford to have a company drop my source," he explained.

Seed companies and farmers are not alone in their rising to meet the need for changes in the seed world. In 2003, more than 70 university breeders, representatives from farmer-based NGOs, and policy specialists met at the "Summit for Seeds and Breeds for the 21st Century" in Washington, D.C. The breeders who represented a diverse set of crops, from schools that have traditionally served clients of large-scale Green Revolution style agriculture, called on each other to "reinvigorate public breeding" to meet the needs of organic and sustainable agriculture.

While the seed consolidation trend seems to have reached a particularly dark moment in the Monsanto-Seminis acquisition, it may also serve to fuel a momentum in the developing organic seed community. The issues are immediate and fraught with complexity; the answers are long-term and require commitment. Said C.R. Lawn of Fedco, "We need to keep working on creating a farmer breeding community. But this is long, slow work. You find and replace varieties one by one. You work slowly in the direction you want to go. Seed work is slow work.

SEED GERMINATION DATABASE

The following data is provided by Thompson and Morgan Successful Seed Raising Guide. This guide is out of print. A seed is an embryo plant and contains within itself virtually all the materials and energy to start off a new plant. To get the most from one's seeds it is needful to understand a little about their needs, so that just the right conditions can be given for successful growth. One of the most usual causes of failures with seed is sowing too deeply; a seed has only enough food within itself for a limited period of growth and a tiny seed sown too deeply soon expends that energy and dies before it can reach the surface.

Our seed guide therefore states the optimum depth at which each type of seed should be sown. Another common cause is watering. Seeds need a supply of moisture and air in the soil around them. Keeping the soil too wet drives out the air and the seed quickly rots, whereas insufficient water causes the tender seedling to dry out and die. We can thoroughly recommend the Polythene bag method which helps to overcome this problem. Watering of containers of very small seeds should always be done from below, allowing the water to creep up until the surface glistens. Most seeds will of course only germinate between certain temperatures. Too low and the seed takes up water but cannot germinate and therefore rots, too high and growth within the seed is prevented.

Fortunately most seeds are tolerant of a wide range of temperatures but it is wise to try to maintain a steady, not fluctuating temperature, at around the figure we have recommended in our guide. Once several of the seeds start to germinate the temperatures can be reduced by about 5 degrees F and ventilation and light should be given. Some perennials and tree and shrub seeds can be very slow and erratic in germination. This may sometimes be due to seed dormancy, a condition which prevents the seed from germinating even when it is perfectly healthy and all conditions for germination are at optimum. The natural method is to sow the seeds out of doors somewhere where they will be sheltered from extremes of climate, predators, etc. and leave them until they emerge, which may be two or three seasons later.

HINTS ON SEED RAISING

Strelitzia and Similar

Do not chip or mark the seedcoat at all but merely remove the orange tuft and soak for up to 2 hours, or even overnight. Sow the seeds in moist sand, pressing them into the sand until only a small part of the black seed is visible and grow in a temperature of 75 degrees F in the dark and ensure that the sand always remains moist. From 7 days onwards inspect the container once a week and as soon as any bulges, roots or shoots are seen remove the germinated seed and pot up in a compost of half peat and half sand. We find that Strelitzias often produce a root without a shoot and we have also found that the young shoots and roots are susceptible to fungal attack. Therefore as soon as possible pot up and provide light and fresh air. Germination can start within 7 days and carry on for 6 months or more.

Palms; Banana; Coffee; Mini-Orange; Tea; Cycads and Similar

All these items can take several months to germinate and are very erratic in germination. Soak for at least 2 hours in warm water before sowing. (After soaking the parchment shell on the Coffee seeds should be removed with the fingernail). Sow in Levington or Arthur Bowers compost and place in the dark in a temperature of 75 degrees F, keeping the compost moist at all times, but not wet. Inspect regularly and occasionally dig around in the compost with a penknife. We normally sow our seeds just below the surface of the soil and we have found that sometimes they make a very vigourous root without producing a shoot at all. If you find a seed with a root then it should be excavated and potted up into a 3-4" pot immediately when it will produce a shoot. Cycads prefer to be potted up into a compost of half sand and half peat. The Tea requires the above treatment but in a lower temperature of 60-65 degree F.

Clivia and Similar

Sow these seeds immediately on receipt in Levington or a peat based compost, covering with a 1/2 " compost. Water and place in the dark in a temperature of 65-70 °F. Germination should occur within 3 weeks.

Ferns (Garden and Indoor)

The fern spore needs a fine film of moisture over which to swim in order to complete the process of reproduction, therefore a good peat compost, such as Levington, ought to be used pressed down very firmly and which is a lot more moist than one would normally have it in order to provide the moisture film. The spore (seed) should be sprinkled close together on the surface of the soil and not covered and the container should be covered with a piece of glass and placed in diffused light, but not darkness. It is essential to ensure that the compost remains moist at all times. Germination which commences with the appearance of a film of green jelly over the soil can take anything from 1 -5 months. You may wish to try germinating the fern spore on blotting paper which is placed in a saucer and kept moist at all times. A transparent cover is inverted over the saucer and the whole lot placed in a well lit but not sunny position.

You can actually see the fern spores developing and when you can see small plantlettes appearing along the jelly the blotting paper should be lifted and placed on the surface of a container of Levington compost and watered well. It should then be covered with a transparent cover which can remain there until the plants are quite large.

Bromeliads; Cineraria

Calceolaria; Insect Eaters (Drosera, Nepenthes, Sarracenias); Living Stones; Meconopsis; Rubber Plants; Saintpaulia; Streptocarpus; Tibouchina; Xmas Cactus; Begonia and similar These seeds should be sown on the surface of the compost and not covered. The compost should be quite moist and we would recommend that you cover the seed container with a piece of glass or clear plastic and leave in a temperature of approximately 65 degrees F in a position which receives diffused light. Once some of the seeds have germinated air should be admitted gradually otherwise the seedlings may damp off. Alternatively the seeds can be sown on to moist blotting paper or kitchen towel placed in a saucer. Cover with a transparent cover and place on a windowsill which receives plenty of light, but not direct sunlight. Keep the blotting paper wet at all times and when the tiny seedlings are large enough to handle prick out into small pots. If the INSECT EATERS are sown using the first method described the compost requires to be both moist yet free draining. Use only pure peat with no fertilizers added to which sphagnum moss should be added if available.

Alstroemeria; Bonsai; Clematis; Hardy Cyclamen; Eucalyptus; Flower Lawn; Helleborus; Hosta; Primula; Iris and similar

Sowing October-February

Sow the seeds in John Innes seed compost, covering them with a thin layer of compost. After watering place the seed container outside against a North wall or in a cold frame, making sure they are protected against mice, and leave them there until the spring. The compost should be kept moist but not wet at all times, and if the seed containers are out in the open then some shelter has to be given against excessive rain. In the spring bring the seed containers into the greenhouse, or indoors on to a well lit but not sunny windowsill and keep the compost moist. This should trigger off germination. If the seeds do not germinate in the spring keep them in cool moist conditions throughout the summer. As each seed germinates we would recommend that you transplant it almost immediately into its own pot.

Sowing March-September

Sow in John Innes seed compost, or something similar, and place each container in a polythene bag and put into the refrigerator (not the freezer compartment) for 2-3 weeks. After this time place the containers outside in a cold frame or plunge them up to the rims in a shady part of the garden border and cover with glass or clear plastic. Some of the seeds may germinate during the spring and summer and these should be transplanted when large enough to handle. The remainder of the seeds may lay dormant until next spring.

Germination of some items, particularly Alstroemeria, Clematis, Hardy Cyclamen and Christmas Rose (Helleborus) may take take 18 months or more. An alternative method for growing PRIMULAS is to sow in a peat based compost which has already been moistened and do not cover the seed. Cover the container with a piece of glass or plastic and grow in the dark in a steady temperature of 60 °F. This is quite adequate and over 65 °F germination will be inhibited. When the seeds start to germinate sprinkle a thin layer of fine compost over them and when the seed leaves come through this, move the box to a well lit place with a temperature of 55 °F.

At no time should the seed box be in full sun. Hardy Cyclamen have been found to germinate best in total darkness at around 55-60 °F. We have had good results with the following method. Place the seeds between two pieces of damp filter paper, Kleenex tissue, etc., then put into a polythene bag and place this into an opaque container in order to exclude all light. Inspect the seeds after a month and remove and prick out as the seedlings appear, returning the ungerminated seeds to total darkness.

Freesia

Soak the seeds for 24 hours and sow in Levington compost, or something similar, and place in a temperature of 50-60 °F. Germination can sometimes be slow.

Nertera Granadensis (Bead Plant)

We recently found that this subject requires a well drained compost which is completely free from fertilizer (*e.g.* moss peat and sand in equal parts). Sow by barely covering the seed and place a sheet of glass over the container, and leave in a temperature of 65-75 °F.

Turn the glass daily as excessive condensation can kill the young seedlings. On germination the seedlings look very thin and spindly and the glass should be removed almost immediately and the seed container moved to a well lit but not sunny position.

Cactus and Similar

Make very shallow furrows in compost with a plant label and sow in these. No seed should be completely buried. Water from beneath and cover with glass and brown paper or black Polythene. Place in a dark position in a temperature of 70-75 °F and keep moist. On germinating move to a light but not sunny windowsill, give plenty of ventilation and water from beneath.

Pot up when they begin to overcrowd. During the first winter only keep warm and do not allow to get too dry. If it is not possible to grow warm then keep them drier. Subsequent years keep relatively dry through the winter. Can be planted outside, plunged to the rim, all summer if required.

Lilies

Successful germination of seeds of some lilies requires a period of warmth followed by one of cold. Method 1. Put seeds in a screw top jar in moist (not wet) peat and keep at 70-75 °F for 3-4 months. Inspect regularly, any normal seedlings (that is having root and seedling leaves) should be pricked out as they germinate. Any seeds which produce roots but not seedling leaves, sow in a pan and keep at 32-40 °F for 3 months. Seed leaves and normal growth will follow.

Method. Sow in a pan in summer (warm spell); put in a frame (or outside covered by a piece of glass) for the winter. Seeds will germinate in spring. Soil Humus rich (peat or leafmould) lime free and very free drainage (use 1/3 grit). Never overwater, keep bulbs almost dry from November to March.

Delicate Seeds

A method which has proved useful for not only small delicate seeds but for a wide range of types is the Polythene bag method. The seeds should be sown on the surface of the moist compost, covered to their recommended depth if necessary and the container is then placed inside a Polythene bag after which the end is sealed with an elastic band. The bag should 'fog-up' with condensation within 24 hours and if this does not occur place the container almost up to its rim in moisture until the soil surface glistens, then replace in the bag and reseal.

The bag is not removed and normally no more watering is required until the seeds germinate. However, it is wise, if left for a long period to check the compost occasionally. The seed container, bag etc. should be placed in a well lit place with a steady temperature. As soon as a fair number of the seedlings emerge remove the polythene bag, lower the temperature a few degrees and provide plenty of light, but not bright sunshine, to ensure that sturdy seedlings develop. It is also helpful to spray the seedlings occasionally for the first 14 days.

SPECIAL TREATMENT

Hard Seeds-Chipping

Some seeds, *e.g.* Sweet peas, lpomaea etc., have hard seed coats which prevent moisture being absorbed by the seed. All that is needed is for the outer surface to be scratched or abraided to allow water to pass through.

This can be achieved by chipping the seed with a sharp knife at a part furthest away from the 'eye', by rubbing lightly with sandpaper or with very small seed pricking carefully once with a needle etc.

Hard Seeds-Soaking

Soaking is beneficial in two ways; it can soften a hard seed coat and also leach out any chemical inhibitors in the seed which may prevent germination. 24 hours in water which starts off hand hot is usually sufficient. If soaking for longer the water should be changed daily. Seeds of some species (*e.g.* Cytisus, Caragana, Clianthus) swell up when they are soaked. If some seeds of a batch do swell within 24 hours they should be planted immediately and the remainder pricked gently with a pin and returned to soak. As each seed swells it should be removed and sown before it has time to dry out.

Stratification (Cold Treatment)

Some seeds need a period of moisture and cold after harvest before they will germinate-usually this is necessary to either allow the embryo to mature or to break dormancy. This period can be artificially stimulated by placing the moistened seed in a refrigerator for a certain period of time (usually 3-5 weeks at around 41 °F).

With tiny seeds it is best to sow them on moistened compost, seal the container in a Polythene bag and leave everything in the refrigerator for the recommended period. However, larger seeds can be mixed with 2-3 times their volume of damp peat, placed direct into a Polythene bag which is sealed and placed in the refrigerator. Look at seeds from time to time. The seeds must be moist whilst being pre-chilled, but it doesn't usually benefit them to be actually in water or at temperatures below freezing. Light also seems to be beneficial after prechilling and so pre-chilled seeds should have only the lightest covering of compost over them, if any is required, and the seed trays etc. should be in the light and not covered with brown paper etc.

Double Dormancy

Some seeds have a combination of dormancy's and each one has to be broken in turn and in the right sequence before germination can take place; for example, some Lilies, Tree paeonies, Taxus need a three month warm period (68-8 °'F) during which the root develops and then a three month chilling to break dormancy of the shoots, before the seedling actually emerges. Trillium needs a three month chill followed by three months of warmth and then a further three month chill before it will germinate.

Outdoor Treatment

The above mentioned methods (12-15) accelerate the germination process and help to prevent seeds being lost due to external hazards (mice, disease, etc.) but outdoor sowing is just as effective albeit longer. The seeds are best sown in containers of free draining compost and placed in a cold frame or plunged up to their rim outdoors in a shaded part of the garden, preferably on the north side of the house avoiding cold drying winds and strong sun.

Recent tests show that much of the beneficial effects of pre-chilling are lost if the seed is not exposed to light immediately afterwards. We therefore recommend sowing the seeds very close to the surface of the soil and covering the container with a sheet of glass. An alternative method especially with larger seeds, is to sow the seed in a well prepared ground, cover with a jam jar and press this down well into the soil so that the seeds are enclosed and safe from predators, drying out etc.

Germination Days

The usual time period in which a particular variety will germinate given optimum conditions.

Light/Dark

Seeds needing light should have no newspaper, brown paper etc. placed over the trays. Seeds needing dark for germination should be placed in total darkness.

Slow and Irregular Germination

This is the column with the "X". Not all seeds will show at once -prick out each seedling as it becomes large enough to handle and don't discard the container until well over the time suggested.

Temperature

A steady temperature between these limits is recommended-fluctuating temperatures can damage a seedling in its critical early stages.

Compost

Most reputable seed composts will be quite adequate and we have indicated where a loam based type such as John Innes or a peat based type such as Levington would be slightly more suitable. On no account should potting composts, which have additional fertilizers, be used.

Sowing Depth

If in doubt sow shallowly, but always ensure that the compost surface is damp. J.C. = Just cover the seed with compost or sharp sand. S = Sow on the surface and do not cover at all with compost.

Sowing in Situ

Where recommended under the heading of comments, these seeds can be sown out of doors. Moist soil worked down to a fine tilth is essential. For hardy annuals and perennials sowing can be carried out from late winter onwards as soon as the ground is workable and has warmed up and half hardy annuals after all danger of frost is passed.

GERMINATION BEGINNING OF GROWTH

Germination is the beginning of growth of a seed. The seed must have the right level of warmth and moisture to begin to germinate. First, the seed leaves absorb moisture which allows the food reserves to become available to the new plant.

It can then produce a root so that it can find its own water, followed by a shoot which develops from the plumule, which will allow it to absorb light. The plant needs both water and light to grow. Sometimes, the seed leaves, or cotyledons, remain below the surface of the soil, as in germination of a Pea, below. This is called Hypogeal Germination.

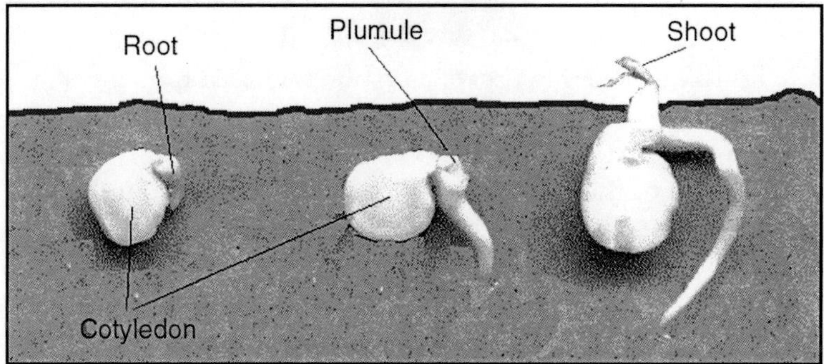

In some species, the seed leaves remain on the new shoot and are brought above the ground, as in germination of the Ash tree, below. This is called Epigeal Germination.

There is one other type of germination that is found only in a few tropical plants.

In this case, one special type of leaf is formed above the ground. This special leaf is not a cotyledon, which remains under the ground. The first leaf is a round dissected or frilly circular leaf with a central stalk like an umbrella, and its purpose is to protect the proper shoot from being trampled on or otherwise damaged before the new plant has enough resources to replace it.

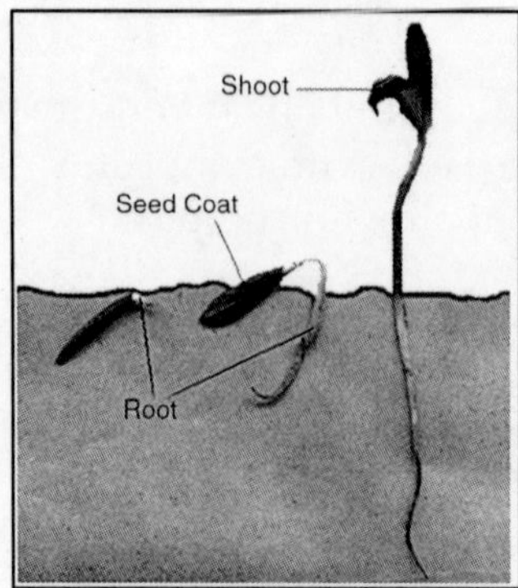

As it is green, it also gathers food by photosynthesis, which enables the root and proper shoot to grow under the ground, so that the new plant is already established when the real shoot emerges. Because the development of the shoot can not be seen, this type of germination is called Cleistogeal Germination from the Greek*cleisto-*, which means hidden.

Fig. This is the Frilly Leaf of a Combretum Seed from Zambia.

GERMINATION

When the pod of the bean is developing, the embryo in the seed is being fed by the parent and visibly grows until ripeness is attained. The young plant then assumes a dormant or resting state within the seed without showing any signs of life. Under certain conditions, however, the plantlet begins to wake up, and soon escapes from its protective coat to lead a separate and independent life. This awakening from a resting condition to a state of active growth is called germination and is dependent upon an adequate supply of

- Heat
- Water
- Air or oxygen.

It is also essential, of course, that the plantlet in the seed must be alive. The exact nature of the dormant state of seeds is not understood, but in old seeds and those which are gathered in an immature condition or badly stored the embryo is often weakened or actually dead; in the latter case no germination is possible. The exact length of time which seeds may be kept before death of the embryo takes place has never been satisfactorily determined; it varies with the species of the seed, its ripeness and composition, and also with the method of storage. In the case of most farm and garden seeds kept in the ordinary way, few of them are found capable of growth after ten years, and a large number die in two or three years.

For present purposes it will suffice merely to mention that age is a determining factor in the germination of seeds. Water is necessary is well known, as beans may be kept indefinitely in a sack or drawer at various temperatures and with access to air without germination taking place.

When placed in moist ground, or between damp blotting-paper, they absorb water very readily. This is most easily observed when beans are soaked for twelve hours in a dish containing water. The water is transmitted through all parts of the coat, but much more quickly and easily through the micropyle and the line of softer material which runs the whole length of the centre of the hilum.

It is rapidly brought into contact with the part of the embryo which grows first, namely, the radicle. The soft spongy thicker part of the inside of the testa lying beneath the hilum stores up a considerable amount of water for the benefit of the developing plant, and the whole of the embryo and the seed-coat absorb water and become softer and larger in consequence; it is only after this swelling has happened that a bean begins to show any signs of germination.

The need of an adequate temperature for germination is a matter of common knowledge among those accustomed to sow seeds. If soaked beans are placed in the ground in midwinter they show little or no signs of waking from their dormant condition, yet when placed under a glass on damp blotting-paper indoors, the radicle makes its exit from the seed in a few days.

Seeds differ in the temperature which is necessary to induce them to germinate, the embryos in some commence to extend their radicles and push their way through the seed-coat even if just kept above freezing point; others require a temperature of 9° or 10°C to start growth. If attempts are made to grow beans at 45°C it will be found to be too hot, and they make little or no progress. Between this high temperature at which growth appears impossible, and the freezing point where the development of the embryo of the bean is also suspended, there is a temperature at which the embryo makes the most rapid progress, and emerges from the seed-coat in the shortest possible time; this most favourable temperature is about 28°C, both above it and below it the germination of the bean is retarded. The supply of fresh air is also an essential condition for growth of the young plant from the bean seed, but the evidence for its need is not so manifest or so generally recognised as the necessity for moisture and warmth.

It will be found, however, when beans are placed in a flask or bottle containing carbon dioxide or hydrogen gas they refuse to germinate, even when they are supplied with a proper amount of water. The peculiar extension or growth of the parts of the interior of the bean seed, and the fact that a suitable supply of water, air and heat is necessary for the manifestation of these changes, suggests to us that we are dealing with a living structure.

This becomes all the more«apparent when we observe that the oxygen of the air is absorbed, and in its place carbon dioxide is given off into the surrounding air, for this is what happens in the breathing of a living animal. Carbon dioxide is produced when beans germinate.

Place twenty soaked beans in a wide-mouthed bottle, and cork them up after showing that a match burns freely in the bottle. Leave them in a warm place for twenty-four hours, and try if a match will now burn in the bottle. The carbon dioxide gas can be poured out into a beaker containing lime water; on shaking, its presence is proved by the lime water becoming 1 milky ' owing to the precipitation of carbonate of lime. The particular use of the water, heat and air to the plant we cannot at present discuss.

Without water the embryo would have little chance of becoming free from the tough and hard seed-coat surrounding it; water softens the latter, and makes it more easily torn by the extending radicle and plumule. In the early stages of the life of the bean plant, from the commencement of germination up to the time when the first green leaves are unfolded, the development and building up of the elongating rootlet and shoot depend upon the thick cotyledons.

At first the latter are thick and fleshy, but as the radicle and plumule grow the cotyledons become softer and thinner, ultimately shrivelling considerably. The cotyledons are leaves, the interior of which is packed with food for the rest of the growing embryo, and a large amount of the water absorbed by the seed is used for the purpose of dissolving the nutrient material in them, and carrying it from them to the various parts of the root and shoot of the young plant where growth is going on the cotyledons are esential to the development of the root and shoot of the embryo by cutting them off as soon as the two latter parts have emerged from the seed-coat. Try separating one cotyledon and then two at various stages of development, and see if the axis (root and shoot) can be made to develop without them. The growth should be allowed to continue some time in order to obtain well-marked effects. Not only do the changes observed innhe embryo of a germinating bean point to the conclusion that it is a living structure and like an animal dependent on a proper supply of water, heat and air for the manifestation of its life, but the parts of a young bean plant after emerging from the seed soon give evidence of the possession of peculiarities which are associated with life.

When put in the ground, the radicle, in coming out of the seed, turns straight downwards and continues to grow in this direction. This is the case no matter in what position the seed is placed. If, after germination has commenced, it is taken and replanted with the primary root pointing to the surface of the soil, the tip of the root soon begins to curve downwards again, and will maintain this course until again disturbed. The plumule behaves in exactly the opposite manner; after emerging from the seed-coat its bent tip grows upwards and away from the root; if the seed is reversed and replanted the plumule begins to curve in such a manner that its tip is driven upwards towards the surface of the ground.

That these peculiarities are somehow connected with life is clear, as dead embryos show no such behaviour. Sow soaked beans in a flower pot or box filled with ordinary garden soil placing them in various positions in it, some laid on the flat side, some with the hilum directed upwards, and others with the hilum downwards. Allow them to grow in a warm place: take them up as soon as signs of germination are noticed, and observe the direction the root and shoot have taken. The peculiar tendency for the root always to go downwards and stem upwards can be investigated by sowing beans in ordinary garden soil and afterwards reversing them.

To avoid error all should be taken up, and then placed again in the soil in various positions—some as they were, a few with their roots and stems reversed, and others laid in a horizontal position. They may be re-examined at the end of a week. When the roots have extended about half an inch take two seeds and suspend them by means of thread side by side in a bottle with their roots downwards and stem upwards. The bottle should contain a little water to keep the air damp.

When the roots have grown about two inches reverse one of the seeds so that its root points upwards tad stem downwards. Notice that the tip of the root of the reversed seed in about twelve hours begins to turn downwards, while the plumule more slowly bends in such a way as to assume the position it had before it was reversed.

The bottle should be placed in a dark box or cupboard to avoid the influence of light on the plant, and fresh air should be blown into the bottle twice a day. Although seeds vary almost indefinitely in regard to size and shape they are similar to the bean in so far as they all contain a young plant packed away within the seed-coats.

In this essential feature all seeds agree with few exceptions, and it is on account of the existence of a young plant within them that they are of use in the raising of crops or plants. The manner, however, in which the embryo is arranged, and the relative size and appearance of its various parts, differ considerably in seeds; moreover, the growth during and after germination is not the same in all cases. A few of the more important and common variations in these respects must be noticed.

GERMINATION OF WHEAT GRAIN

A wheat grain, which may be taken as an example, is not a seed, but a kind of nut with a single seed within it. The seed grows in such a way as to completely fill up the interior of the nut, and become practically united with its inside wall. The embryo occupies only a small part of the grain, the rest being taken up by the floury endosperm of the seed. The embryo is easily seen at the base of a soaked grain on the side opposite the furrow.

When removed it has the appearance. The part of it which lies close up to the endosperm is a flattened somewhat fleshy shield-shaped structure called the scutcllum attached to the front of the scutellum is the plumule, consisting of a bud formed of an extremely short stem, upon which are sheath-like leaves enclosing each other. The embryo generally possesses five roots, one primary and two pairs of secondary. They are all completely enclosed by a sheath which continuous wjtn the scutellum, and are consequently not visible from outside; their position, however, is marked by projecting bosses. The sheath round the roots is termed the coleorhiza, and when germination takes place it expands and bursts the coats of the grain, the roots about the same time breaking through the enclosing coleo-rhiza.

When a wheat grain is sown in the ground it remains there, but the plumule grows upwards, its first leaf, the coleoptile, appearing above the soil as a single pale tube-like structure; from a slit in the tip of the latter the first flat green blade soon appears, and is followed by a succession of single green leaves, the younger ones growing from within the older ones in regular order. With a sharp knife or razor cut through from back to front, so as to divide the grain into two longitudinal halves, and note the floury endosperm and the shape and parts of the divided embryo. Place a folded sheet of damp blotting paper on a plate, sow some soaked wheat grains on it, and cover with a tumbler. The grains will germinate. Watch their development up to the time the first green leaf appears, taking out the embryo and examining it at different stages of its growth. There is difference of opinion as to which part of the embryo is to be considered the cotyledon.

Soine authorities regard the scutellum as the cotyledon, while others give this name to the coleoptile or first sheathing leaf which comes above ground, and which has no green blade. Others, again, consider that the first sheathing leaf is an extension of the scutellum, and the two combined is therefore the cotyledon. In any case, there is only one cotyledon present, and wheat therefore belongs to the class of monocotyledonous plants. During the growth of the embryo of a wheat grain, it will be noticed that the endosperm becomes soft and decreases in quantity as the roots and plumule expand and develop; the endosperm is the food upon which the young plant depends during the early staggs of its life, the scutellum acting as a structure for changing, absorbing, and transferring this reserve-food to the growing parts which need it. Remove the embryos from well-soaked grains, and grow them without the endosperm on damp blotting-paper.

Allow ordinary uninjured grains to grow with them. Both the embryos in the grains and those removed from the grains develop, but there is a great difference in the results after a few days. The store of reserve-food on which the young plant depends for its early development is sufficient to enable it to form a root, stem, and several leaves, as is evident when seeds are allowed to germinate upon damp flannel or blotting-paper, from which nothing but water is absorbed. No food-materials or manures are needed for this primary development, and seeds germinate and the seedlings grow for a considerable time as well in poor soil or sand as in good rich ground.

As soon as the reserve store is exhausted hunger becomes apparent, and unless the plants are then supplied with suitable nutriment from the soil and air, and are also placed under conditions favourable for growth, weakness and death are likely to occur. Among the larger seeds, such as beans and peas, where there is an abundant store of reserve-food, the young seedlings begin to manufacture food for themselves from materials absorbed from the soil and air, long before their reserve is exhausted.

In small seeds the reserve is sometimes almost consumed before the roots and leaves are sufficiently developed to carry on their work properly, in which cases a more or less temporary starvation and check to growth ensues. Especially does this happen when seeds are sown too deeply, for a large amount of food is then used in the production of a stem long enough to lift the leaves up into the air.

SEED OF WHITE MUSTARD

The seed of white mustard (Brassica alba Boiss.) contains an embryo which like that of the bean consists of a radicle, plumule and two cotyledons; the latter, which are folded together, are relatively thinner than those of the bean and deeply notched. The radicle is bent round and lies in the fold of the cotyledons, between which is the very small, almost invisible plumule. On germination the cotyledons, instead of remaining within the seed-coat and below the ground as in the case of the broad bean, escape from the enclosing coats altogether and grow up out of the ground, enlarging at the same time, and becoming green like ordinary leaves. They are the first ' smooth' leaves of the seedling mustard plant. After a short time the plumule grows up from between the cotyledons and forms a stem upon which are gradually unfolded the ordinary divided ' rough' leaves. Soak white mustard seeds, and examine their structure, noting especially the way in which the embryos are packet in them.

Allow some to germinate and grow for a week or more on damp flannel, and examine them in various stages of development, noting the notched cotyledons with small. The term hypogean is applied to cotyledons which remain below ground, those coming above being epigean, the relative amount of growth in the hypocotyl and epicotyl determining their position. If the hypocotyl grows vigorously during or after germination the cotyledons are forced above ground; when only the epicotyl grows the plumule is lifted up above the soil, but the cotyledons remain below where the seed is placed. In the broad bean the hypocotyl is very short, and the point where it ends and the root begins is not clearly denned. In a mustard seedling, however, the point of separation between the root and stem is somewhat swollen and readily distinguished.

All plants whose embryos, like those of the bean and mustard, possess two cotyledons, are known as Dicotyledons they form a very large, well-marked class of the flowering or seed-bearing plants. The seeds contain within their coats nothing but an embryo plant, which depends for the development of its root and shoot upon the substances stored up in some part of its body, its cotyledons chiefly.

This is true even in the case of seeds like those of white mustard, in which the cotyledons of the embryo are comparatively thin. There are, however, a number of plants, such as the ash, mangel and potato, which, although belonging to the Dicotyledons, have seeds in which there are stores of food inside the seed-coat, but free from the embryo and its cotyledons. Such separate reserve-food is stored in that part of the seed known as the endosperm or 'albumen,1 and seeds in which it is present are called endospermous or albuminous seeds. Those like the bean, pea, and vetch, mustard, and turnip, which have no separate reserve-food, are known as exendospermous or exalbuminous seeds. Take out a seed from the fruit of the ash tree in autumn; carefully cut thin shavings from the flat side of the seed, stafting about the middle of the seed and cutting towards the narrow end. Note the white embryo with its well-marked radicle, hypocotyl and two flat cotyledons lying within the semi-transparent endosperm. Some of the most commonly occurring endospermous seeds will be found to have embryos within them which are not dicotyledonous, and whose structure is in many respects. A good example is met with in the onion.

ONION SEED

The seed is black, somewhat oval in outline, with one side convex, the other almost flat. Each contains within it endosperm and an embryo which lies curled up inside in the form seen. When germination commences, the curved part imbedded in the middle of the endosperm grows and forces the end of the embryo out of the seed. From this exposed end, which is the radicle, a straight, slender, primary root develops. The part of the young seedling which extends from the root into the interior of the seed, grows very rapidly at first, at the same time assuming a sharply bent outline.

It comes above ground in the form of a close loop but on further growth the end within the seed is pulled out of the soil, and grows up in the air. The tip within the seed changes and absorbs the endosperm, and usually remains there until all the nutrient material has been transferred from it to the various centres of growth in the young plant. After the food-reserve is exhausted the tip withers and becomes free from the seed coat.

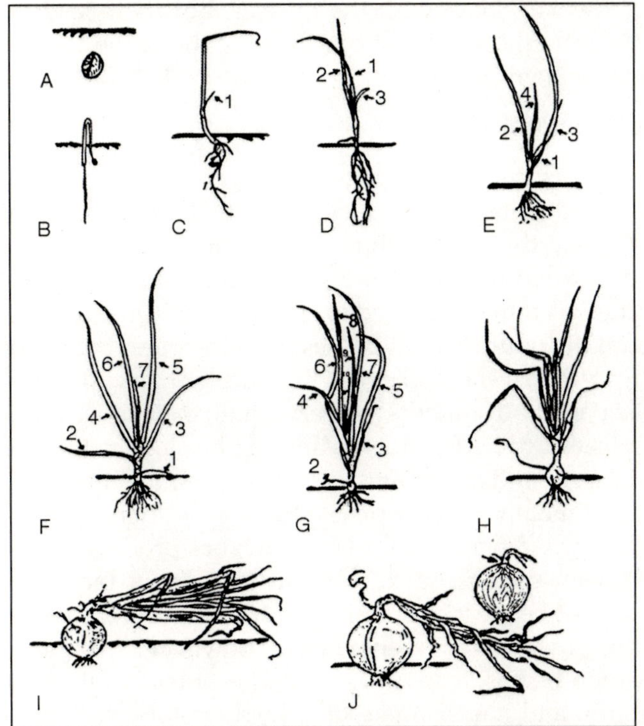

Fig. Germination of Onion

In loose soils the latter is pulled above ground before the endosperm is exhausted, and remains on the end of the tip for some time. In other cases where the soil is damper and of a stiffer nature the seed-coat remains below ground altogether. The curved part of the embryo which comes above ground is a leaf. It is the cotyledon of the embryo, and is in reality a thin hollow leaf like those of the full grown onion plant: within it is the plumule, which consists of a series of hollow conical leaves arranged one inside the other. After one leaf emerges others soon follow, the younger ones coming out in regular order through slits in the sides of those immediately older than themselves. Soak fresh onion seeds in water for a few hours. With a razor cut through some parallel to their flat sides in order to show the embryo within.

Sow others in damp blotting-paper; allow them to germinate and the seedlings to develop; make observations of them at different stages of growth. Watch the germination of seeds sown in boxes or pots containing ordinary garden soil. Plants whose embryos possess only one cotyledon are known as Monocotyledons, and form the second large class of seed-bearing plants. Few of the representatives of this class with which we are ordinarily familiar have true seeds large enough for examination.

The onion is probably one of the best commonly occurring examples which may be considered typical of the monocotyledons and easily obtainable.

To this class, however, belong all the grasses, but their seeds and embryos are so different in many respects from those of the onion that it is necessary to examine one of them in detail.

The Common Bean of Seeds

A broad bean is one of the largest seeds met with in ordinary farm or garden practice, and as its parts are all sufficiently large to be observed without the special aid of anything more than an ordinary pocket lens, it is especially fitted for study. When a nearly ripe pod of a broad bean plant is opened, each seed within it is found attached to. The inside by means and it is through this stalk that all the nourishment passes from the parent to enable the young seed to develop. A1 first the pod exists in a rudimentary form in the centre of a flower and its parts and contents are very small; they are nevertheless readily seen with a pocket lens. After the fading of the flower, the pod and seeds within it grow larger and larger at the expense of food supplied by the rest of the plant, and ultimately when ripe the funicles wither and dry up, and the seeds become detached from the parent which has produced them. When dry and ripe each bean seed is hard, with an uneven surface, but its internal construction cannot be clearly examined in this condition.

On soaking in water for twelve hours, however, it becomes softer, and the parts can then be easily investigated. The outside, which is a pale buff colour, is smooth, and has at one end a narrow elongated black scar called the hilum of the seed. It is known popularly as the ' eye' of the bean, and marks the place where the broad end of the funicle separated from the seed when it ripened in the pod. Quite close to one end of the hilum is a very minute hole known as the micropyh, easily seen with a lens, and through which water oozes out usually accompanied by bubbles of air when soaked beans are squeezed between the finger and thumb. This opening communicates with the interior of the seed, and is the only one it possesses. The rest of the seed after the testa is removed, is of oval flattened shape similar to the complete bean, and is divisible into two large fleshy halves called cotyledons which, however, are not completely separate from each other, but connected at the side with a conical projecting body, one end of which is found to fit into a hollow cavity in the seed-coat exactly opposite the micropyle; the other end is bent and turned inwards between the fleshy cotyledons.

Small Curved Structure

The extent and shape of this small curved structure is most easily observed when one of the cotyledons is removed completely; it remains attached to the other. Soak some broad beans in water and keep them in a warm place all night. Examine them next day and make drawings of the various parts seen both before and after stripping off the testa. Observe the relative position of the parts of the embryo in reference to each other and to the seed-coat.

Examine and compare the structure of the following seeds after soaking in the same way:—Pea, scarlet runner beans, vetches, and red clover. The bean seed contains nothing more than what has already been described; the nature and relationship of its component parts only become intelligible when the seed is placed in the ground or maintained under certain conditions, and allowed to grow. When growth commences the lower end of the small curved structure elongates and breaks its way through the coat of the seed at a point very close to the micropyle, but not, as often erroneously stated, through the micropyle itself. It soon assumes the form and is recognised as a root of a young bean plant The upper bent half, which lies between the cotyledons, also pushes its way out of the same opening in the seed-coat and develops into a stem, from the tip of which leaves are gradually unfolded.

It is thus seen that the seed of a broad bean is a packet containing a bean plant in a rudimentary condition. This plantlet is callai an embryo, and the portion of it which becomes root and stem is its primary axis. The part of the primary axis which is below the point where the cotyledons are attached consists of a very small piece of stem, the hypocotyl at the end of which is the radicle or primary root. Where the stem ends and the root begins cannot be determined in the bean seedling without the aid of the microscope and examination of the internal structure of the axis of the plant. The curved end of the primary axis above the cotyledons is the plumule of the embryo, and consists of a very short piece of stem, the tpicotyl on the top of which is a bud.

From the latter is derived the ordinary stem which comes above ground with its green leaves and flowers. In the early stages of the growth of the embryo from the seed the hypocotyl grows very little. The upper part of the stem bearing the plumule comes out of the seed bent and it maintain this curved shape for some time after emerging. By this behaviour the delicate leaves of the plumule are protected from injury during their progress upwards when a seed is sown in earth or sand. Fold up some soaked beans in two thicknesses of white flannel made damp, and place them on a plate. Cover them with another plate placed upside down, and leave them in a warm room. Examine them twice a day, leaving them, exposed to the fresh air for a few minutes each time, and keeping the flannel damp, not wet. When they sprout notice the place where the radicle has come out of the seed-coat. Let some grow till the radicle and plumule are well out of the seed, and compare the various parts of the sprouted seeds with unsprouted ones.

PRIMARY AND SECONDARY BOOTS OF SEEDLING

It was noticed, when dealing with the bean seedling, that its two ends always grow in opposite directions; the plantlet can be considered as an elongated axis, one end of which bears the leaves and invariably comes above ground, while the opposite end never bears leaves, and persistently follows the plumb-line downwards.

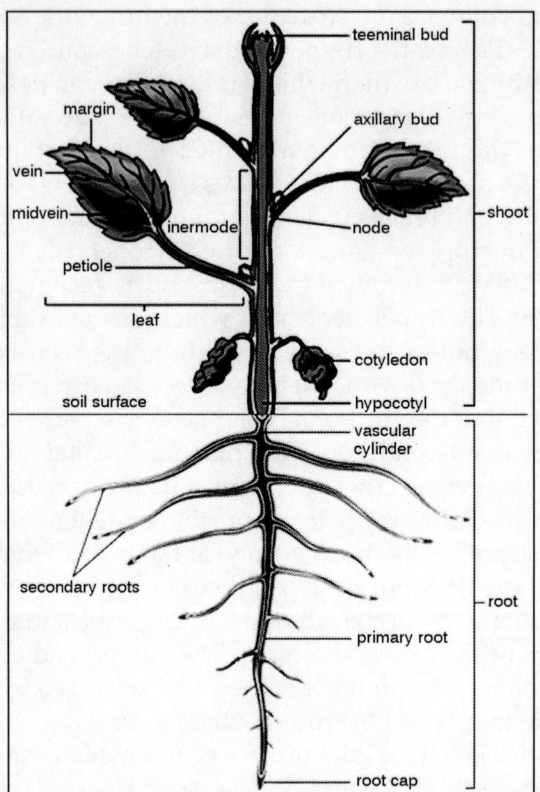

Fig. Dicotyledonous Plants

The descending part is known as the root. As will be pointed out later, all roots do not behave in this manner, and it is to be specially noted that many of the underground parts of plants are not roots; the exceptions, however, may be left for future consideration. The first or primary root which the bean plant possesses is merely an extension of the radicle of the embryo which exists nrithin the seed.

Soon after making its exit from the seed, it takes a downward course, and elongates by growth taking place near its tip. Another point to be noted is that the lateral roots do not arise as up-growths on the surface of the primary root, but come from within it, and are described as endogenous. The slits which they make in the substance of the primary root, and through which they emerge, can be readily seen in a bean seedling.

The secondary lateral roots are connected with its central more solid core; the three lowest ones, although they have just begun to grow, have not yet penetrated the outer layer of the root, and would not be visible on the outside of the latter. This mode of origin is characteristic of lateral roots generally wherever they are met with. Many dicotyledonous plants have roots similar to those of the bean plant.

When, as in this case, the primary one continues to grow, keeping distinctly larger than the lateral ones, it is called a tap root. Very good examples are met with among cultivated plants in the carrot, mangel, red clover, and mustard in shep-herd's-purse, poppy, and many other weeds, as well as in most broad-leaved trees. A number of plants have swollen fleshy roots in which food materials are stored for future use; they are described as tuberous roots, and must be distinguished from tubers, which are fleshy underground stems. To designate the different forms of thickened roots various special terms are in use. The typical carrot root is conical; that of the turnip napiform. The root of the radish is spoken of as fusiform. In some instances the primary root is soon rivalled in size by its branches; it may even cease growth altogether. Such plants, on being pulled out of the ground, exhibit a bunch of slender roots, the chief of which are all much the same diameter and length; roots of this character are described as fibrous and are well exemplified in common groundsel and grasses.

GERMINATING AND GROWING AVOCADOS AS HOUSE PLANT

The Portuguese originally introduced avocados to Sri Lanka, the Sinhalese call it Aligata Pera, giving it the name of Alligator Pear in the USA. It is also known as butter fruit, butter pear and laurel peach. Growing areas include Israel, Mexico, Spain, South Africa and California.

GETTING STARTED

Select an avocado that is soft when you give it a squeeze. Carefully cut fruit in half lengthwise to get to the seed. A gentle twist of the cut fruit will release your seed. Remember whether you grow it in water or in soil, set the seed with its base (the wider portion) down and the tip up. Having trouble telling which end goes down? Look for the slight stem scar on one end. That is the part that goes down and will grow roots.

Method 1

Wash avocado seed to remove any pulp. Using a very sharp knife you can cut off a thin slice of the top and bottom of the seed, this speeds up the germination. Wrap in very damp paper towel. Place in a covered dish and put it in a dark place for 2-4 weeks. Check in now and then to see if anything is happening. The taproot is generally the first growth to merge from the seed. Once the root is around 3 inches long plant your avocado following the potting method.

Method 2

Another method to sprout the seed, remove the large seed from the centre of the fruit and wash it in water. For propagation purposes, the broad end of the seed is considered to be the bottom. The pointed end is the top. Insert several toothpicks into the sides of the seed. They should be placed about halfway up the pit. Then suspend the seed in a glass of water. The bottom 1/4 of the seed should rest in water. During this time add water to maintain the initial water level. Occasionally dump out the water and replace with fresh water. Place on a windowsill in good light.

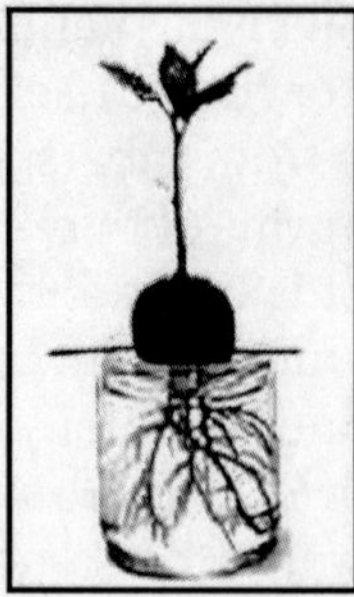

However, do not leave in direct sunlight for long periods. In about two to three weeks, the pit will start to crack. In another three to four weeks, a single root will appear at the submerged end of the pit. In the next one to three weeks a stem will start growing. When the root is two or three inches long and the stem is at least an inch or two you will be ready to plant. If it doesn't sprout within 2 to 3 months, discard the original avocado pit and begin another. As not all pits will grow we recommend starting several pits at one time so you get at least one or more plants.

Method 3

Just plant the pit in the soil from the beginning. Carefully peel off the brown seed coat of a fresh avocado pit and plant the seed in a 6-inch container (with a drainage hole) filled with a rich, well draining potting soil. Leave about 1/of the top of the seed protruding from the soil. Keep the soil moist but not soggy. Do not allow soil to dry out. It can take a month or longer for the avocado sprout to emerge using this method.

Potting for Growing Methods 1 and 2

Pot the seedling when the root system has become well developed; the roots should be at least 2 to 3 inches long. Plant the white taproot in dirt leaving the top "red sprout" and 1/3 of the upper seed exposed. Position the seed in the centre of the pot.

A 6" plastic or clay pot is suitable and please put some gravel in the bottom of the pot for drainage. A good soil mix for avocados is: 1 partCoco Peat, 1 part perlite and 1 part sterilized potting soil. Pack the dirt well around the seed. Keep the soil fairly wet for the first week.

Repotting

Once the plants filled their pots up with healthy roots, they should be potted in larger ones. Repotting should be done in the spring.

GROWING TIPS

Bright indirect light is best for young avocado plants, direct sun will give younger plants a sunburn! They prefer moderate temperatures of 60 to 80 °F. Your plant can grow 4-5 feet tall rather quickly. It's a good practice to mist the leaves of your avocado every few days if the air in your home is very dry. Spring is the best time for pruning. If you would like to prune to keep it bushier pinch just the tip of the growing top when your plant is 1 foot tall. Repeat the procedure when it has grown to 2 to 3 feet tall. This should give you slower growing but very sturdy side branches. They don't always respond well to pruning.

WATERING

Watering avocadoes correctly is critical to their success as a house plant.

Avocados are very susceptible to root rot. Using too much water applied too often means the soil will stay wet for long periods, causing the roots in the waterlogged soil to begin to rot and die. With their roots gone the plants wilt as if they are thirsty.

These plants will not perk up when they're watered. If the soil is kept soggy with water, then it contains no oxygen so the roots won't function and will rot. You want to use pots with drainage holes for the plants. No closed containers please. When you water them the excess water in the saucer should be dumped so it isn't absorbed back into the soil. Again use a potting mix that allows good drainage The soil mix should be light and porous not heavy. Watering from above is recommended. Water should be added to the point that some comes out the drainage holes. Another mistake in watering is to constantly give the plant small amounts of water like an 8 ounce cupful. This will actually allow the rootball to dry out.

FERTILIZING

Fertilize every 2 weeks month during the spring and summer. In winter fertilize just once every 6 weeks. Our Golden Harvest Natural Fertilizer is great for your avocado plant!

PROBLEM SOLVING

- *Yellowing, Dropping Leaves*: Too much water or not enough sunlight can cause yellowing and dropping of the lower leaves. Cut back on water and move to a sunnier location.
- *Browning of Leaf Edges*: This can be caused by a lack of humidity. Setting the pot on a tray of pebbles and water will help increase humidity around your avocado plant but don't let the water level in the tray touch the bottom of the pot. Sometimes the brown leaf edges stay no matter what we do. The brown area can be trimmed off for aesthetic purposes.

- *General Yellowing of new or older Leaves*: This is a sign of fertilizer deficiency. We will see yellowing of the leaf tissue while the veins remain green to tip us off to this situation. Fertilize once or twice a month during the spring and summer. Winter is rest time. Our Golden Harvest Natural Fertilizer is great for your avocado plant!

PALM SEED GERMINATION

Anyone can grow their own palm for almost no cost. This is good news since, in the UK at least, it will cost you a great deal to buy an adult palm. Even the seeds of most varieties are fairly expensive and, unless you've already mastered the basics, you may be unsuccessful in growing your seeds into plants. For these reasons, I recommend starting out by growing your own date palms *(Phoenix dactylifera)*. You can get the seeds of these easily. Just buy yourself a packet of dates from your local supermarket. Eat the dates and keep the stones. Just about any dates will do - either fresh or the more readily available dried ones. Although the dried ones are normally sugar-coated, that doesn't affect their viability. The only dates that aren't suitable are stoned ones or chopped ones - because, obviously, neither of these contains the seeds!

Once you've got your date stones, soak them overnight in a cup of water. Then clean the stones carefully to remove all fragments of date flesh. You can then, if you wish, simply pop them into pots and wait for them to sprout. However, this is a very haphazard method and will probably produce disappointing results. If you want a more reliable method of getting from seed to seedling, try out the following steps:

THE COLLINGBOURNE PALM GERMINATION METHOD

When I first started growing my own palms from seeds I was very disappointed by my poor success rate. Most of the books on palms and on propagation in general were a bit vague on the specifics. So I had no choice but to carry out my own experiments. Soak the seeds for 24 hours (sometimes longer for very hard-coated seeds). Change the water several times during this period. Keep the soaking seeds in a warm place such as a kitchen or airing cupboard.

Clean the seeds. Get rid of any flesh or hairs (if possible) adhering to them. Fungus is a big problem with some germinating palm seeds. Although you can treat them with fungicides, these are not always particularly efficient. Fungus loves the flesh of seeds, so clean it off!

Place the seeds in a sterile medium which is only barely moist. The biggest problems I had when I started growing palms were due to the fact that the germinating medium was too wet. This grew lots of mould, but not many palms. These days, I use a 1 pint measure - about 550ml - of vermiculite (available from garden centres), moistened with 30ml - about two tablespoons - of water. The resulting vermiculite should feel quite dry to the touch.

Nevertheless, don't be tempted to add more water. I have germinated seeds in as little at 10 ml of water to 1 pint of vermiculite. Seal the seeds in an airtight container - such as a plastic sandwich box (recommended) or a plastic bag. Place it in a warm-to-hot position, say above the hot-water tank. Temperatures of 90 degrees Fahrenheit (about 32 °C) are ideal.

Inspect the seeds daily. If the seeds are fresh (a big if!), you may find that many species germinate in as little as one or two weeks, in spite of the fact that the books tell you they are likely to take 2 or 3 months! When a root appears, carefully put the seeds into well-drained compost (*e.g.* 50/50 standard potting compost and coarse grit), and keep warm until the shoot appears.

When the shoot appears place the palm in a warm, light place. If your room temperature is fairly cool, be very careful not to over-water as the roots of some palms (even the humble date palm) will rot if the compost is kept wet when the palms are not actively growing.

GLOBAL SEED MARKET

Despite resistance from several sectors, this market has fiercely penetrated the agriculture sector. The global seed market is projected to grow driven by rising standards in global farming. Farmers in various countries are showing increased preference for certified seeds and innovative high-yielding varieties. Extensive use of biotechnology in seed development is resulting in new and high-end varieties, which is driving the global seed market in value terms. Genetically modified crops, first cultivated in 1996, has rapidly increased penetration in terms of total acreage grown over the past decade.

Genetically modified seeds is expected to grow from the fringes of crop protection into newer potential applications, thus increasing the presence of biotechnology in the industry. The acreage under biotech crops increased to about 333 million in 2009, with about 14 million farmers spread across 25 countries having adopted biotech crops. Currently, the industry is witnessing new lines of genetically modified crops with improved pest protection and nutritional value being added by several producers.

The US represents the single largest market for seeds, as stated by the new market research report on Seeds. Demand for seeds in the United States is expected to maintain steady growth in the coming years, spurred by the development and launch of a wide range of high-end transgenic and hybrid seeds. The market is also likely to be shaped by a number of favourable factors, such as growing garden and lawn expenditures of consumers, a general shift towards narrow planting of rows, and expanding new uses of crops. Europe and Asia-Pacific are the other important seeds markets. Demand for seeds is forecast to be the fastest in Asia-Pacific, which is expected to increase at a compounded annual growth rate of more than 5.0% through 2015.

Segment wise, Grain Seeds represents the largest segment. Horticulture seeds constitute the second largest segment in the seeds market. However, demand is expected to mainly emanate from Vegetable Seeds, which represents the fastest growing segment during the analysis period. The vegetable seeds market is primarily a fragmented and complex sector within the seed industry. The homeowner segment of the vegetable seed market is directly related to personal preferences and values. Major players profiled in the report include Bayer CropScience AG, Limagrain, Monsanto Company, Mycogen Seeds, Pioneer Hi-Bred International, Inc., Sakata Seed Corporation, Svalof Weibull AB, Syngenta Crop Protection, Golden Harvest Seeds Inc., among others.

The research report titled "Seeds: A Global Strategic Business Report" announced by Global Industry Analysts Inc., provides a comprehensive review of the Seeds market, current market trends, key growth drivers, use of biotechnology in agriculture, review of genetically modified (GM) seeds, recent product introductions, recent industry activity, and profiles of major/niche global as well as regional market players. The study analyses market data and analytics in terms of value sales for the global seeds market for the following geographic markets—US, Canada, Japan, Europe, Asia-Pacific, Latin America, and Rest of World. Key segments analysed include Grain Seeds, Vegetable Seeds, Oilseeds, Horticulture Seeds, Fruit Seeds, and Miscellaneous Seeds. A seven-year historic analysis is provided for additional perspective.

SEEDS THREATEN

The 12,000-year-old practice in which farm families save their best seed from one year's harvest for the next season's planting may be coming to an end by the year 2000. In March 1998, Delta ~ Pine Land Co. arid the US Department of Agriculture (USDA) announced they had received a US patent on a new genetic technology designed to prevent unauthorized seed-saving by farmers. The patented technology enables a seed company to genetically alter seed so that the plants that grow from it are sterile; farmers cannot use their seeds. The patent is broad applying to plants and seeds of all species including both transgenic (genetically engineered) and conventionally-bred seeds. The developers of the new technology say that their technique to prevent seed-saving is still in the product development stage, and is now being tested on cotton and tobacco. They hope to have a product on the market sometime after the year 2000.

Over the last four years, USDA researchers claim to have spent nearly $190,000 to support research on what the Rural Advancement Foundation International (RAFI) calls "Terminator" seed technology. Delta and Pine Land, the seed industry collaborator, devoted $275,000 of in-house expenses and contributed an additional $255,000 to the joint research. According to a USDA spokesperson, Delta and Pine Land Co. has the option to exclusively license the jointly developed patented technology. The USDA's Willard Phelps explained that the goal is "to increase the value of proprietary seed owned by US seed companies and to open up new markets in second and third world countries." USDA molecular biologist Melvin J. Oliver, the primary inventor of the technology, explained why the US developed a technology that prohibits farmers from saving seeds: "Our mission is to protect US agriculture and to make us competitive in the face of foreign competition. Without this, there is no way of protecting the patented seed technology." USDA stands to earn royalties of about 5 per cent of the net sales if a product is commercialized. The day after the. patent was announced, Delta and Pine Land Company's stock rose sharply. While USDA and seed industry profits may increase, these earnings come at enormous cost to farmers and to global food security. USDA researchers interviewed by the authors expressed a strong allegiance to the commercial seed industry and an appalling lack of awareness about this technology's potential effects, especially in the US South.

IMPACT

Delta andr Pine Land Co.'s press release claims that its new technology has " the prospect of opening significant worldwide seed markets to the sale of transgenic technology for crops in which seed currently is saved and used in subsequent plantings." Up to 1.4 billion resource-poor farmers in the South depend on farm-saved seed and seeds exchanged with neighbours as their primary seed source. A technology that restricts farmer expertise in selecting seed and developing locally-adapted strains is a threat to food security and agricultural biodiversity, especially for the poor. The threat is real, especially considering that USDA and Delta andr Pine Land have applied for patent protection in countries from Brazil to Vietnam.

If the Terminator technology is widely licensed, it could mean that the commercial seed industry will enter entirely new sectors of the seed market - especially in self-pollinating seeds such as wheat, rice, cotton, soybeans, oats and sorghum. Historically, there has been little commercial interest in non-hybridized seeds such as wheat and rice because there was no way for seed companies to control reproduction. With the patent announcement, the world's two most critical food crops - rice and wheat, staple crops for three-quarters of the world's poor—potentially enter the realm of private monopoly.

In May, Monsanto announced it would acquire Delta andr Pine Land Company for $1.8 billion. This means that seed-sterilizing technology is now in the hands of the world's third-largest seed corporation and second largest agrochemical corporation.

Monsanto's 1996 revenues were $9.26 billion. The company's genetically engineered crops are expected to be used on approximately 50 million acres worldwide in 1998.

If Monsanto's new technology provides a genetic mechanism to prevent farmers from germinating a second generation of seed, then seed companies will gain the biological control over seeds that they have heretofore lacked in non-hybrid crops. Nobody knows exactly how many farmers in industrialized countries save seed from their harvest each year. By some estimates, 20 to 30 per cent of all soybean fields in the US midwest are planted with farmer-saved seed. Most North American wheat farmers rely on farm-saved seeds and return to the commercial market once every four or five years. Almost all of the wheat grown on the Canadian prairies is from seed produced in the communities in which it is grown. The same is true for lentils and peas.

OPTIONS FOR FARMERS

Proponents of the Terminator technology are quick to point out that farmers will not buy seed that does not bring them benefits. But market choices must be examined in the context of privatization of plant breeding and rapid consolidation in the global seed industry. The top ten seed corporations control approximately 40 per cent of the commercial seed market. Current trends in seed industry consolidation, coupled with rapid declines in public sector breeding, mean that farmers are increasingly vulnerable and have far fewer options in the marketplace.

A new technology that is designed to give the seed industry greater control over seeds will ultimately weaken the role. of public breeders and reinforce corporate consolidation in the global seed industry.

Advocates of Terminator technology claim that it will be a boon to food production in the South, because seed companies will have an incentive to invest in crops that have long been ignored by the commercial seed industry. But private companies are not interested in developing plant varieties for poor farmers because they know the farmers can't pay. Existing national public breeding programs tend to focus on seeds for high-yielding, irrigated lands, leaving resource-poor farmers to fend for themselves.

Half the world's farmers. are poor and can't afford to buy seed every season, yet poor farmers grow 15-20 per cent of the world's food and directly feed at least 1.4 billion people - 100 million in Latin America, 300 million in Africa, and one billion in Asia. These farmers depend upon saved seed and their own breeding skills in adapting other varieties for use on their often-marginal lands.

BIOSAFETY CONCERNS

The seed industry is expected to defend the Terminator technology by arguing that it will increase the safety of using genetically-engineered crops. Since the seed carries the sterility trait, say proponents, it is less likely that transgenic material will escape from one crop into related species and wild crop relatives. The seed industry is expected to argue that this built-in safety feature will speed up biotech advances in agriculture and increase productivity. Molecular biologists who have studied the patent have mixed views on the potential ecological hazards of the sterility trait. The greatest fear is that the sterility trait from first generation seed might spread via pollen to neighbouring crops or wild relatives growing nearby. Some biologists argue that pollen even if pollen does escape, it would not pose a threat. The danger is that neighbouring crops could be rendered "sterile" due to cross pollination - wreaking havoc on the surrounding ecosystem. Given that the technology is new and untested on a large scale, biosafety issues remain an important concern.

REACTIONS TO THE TERMINATOR

"This is a patent that is too profitable for companies to ignore," says Camila Montecinos of the Chilean-based Center for Education and Technology. "We will see pressure on national regulatory systems to marginalize saved-seed varieties and clear the way for the Terminator. More than a billion farm families are at risk."

"Governments should declare use of the technology illegal," she insists. "This is an immoral technique that robs farming communities of their age-old right to save seed, and their role as plant breeders."

To this, corporate breeders respond that the new technology simply does for hard-to-hybridize crops what the hybrid technique did for maize. Hybrid seed is either sterile or fails to reproduce the same-quality characteristics in the next generation. Thus, most maize farmers buy seed every year.

"Poor farmers can't afford hybrids either," Montecinos points out, "but there's a key difference. The theory behind hybridization is that it allows breeders to make crosses that couldn't be made otherwise and that are supposed to give the plant higher yields and vigour. The results are often disappointing, but that's the rationale. In the case of Terminator technology, there's absolutely no agronomic benefit for farmers. The sole purpose is to facilitate monopoly control, and the sole beneficiary is agribusiness."

Neth Dano of the civil organization SEARICE, based in The Philippines, sees a threat to the environment and to long-term food security: "We work with farmers who may buy a commercial variety, but its breeder wouldn't recognize it five years later. Women select the best seeds every year, and, over time, the rice molds itself to the farm's ecosystem. Women also cross the commercial variety with other rice strains to breed their own locally-adapted seeds. The Terminator could put an end to all this and increase crop uniformity and vulnerability. It poses a threat to the culture of seed-sharing and exchange that is led primarily by women farmers."

TERMINATE THE TERMINATOR

At the fourth Conference of the Parties (COP) to the Convention on Biological Diversity meeting in Bratislava, Slovakia, May 4-15, 1998, the Philippine resolution calling for a ban on the technology was supported by delegates from Kenya, Zambia, Pakistan, Rwanda and Sri Lanka. When it was announced on May 12th that the Delta and Pine Land Co. had been acquired by Monsanto, concerns were heightened about the potential dangers of this technology for farmers and food security. The COP has requested that the issue be considered by its Subsidiary Body on Scientific, Technical and Technological Advice.

A genetic technology aiming to sterilize seed threatens to extinguish the right of farmers to save seed and breed new crop varieties, and threatens the food security of 1.4 billion people. RAFI and other nongovernmental organizations are calling for a global ban on the use of Terminator seeds. Both the patent and the technology should be rejected on the basis of common sense, food security and agricultural biodiversity.

CONJECTURE AND CONCERN

While news of Monsanto's acquisition of Seminis was less than a blip on the general public's radar, small groups of farmers, activists and seed trade professionals immediately began to connect to discuss the ramifications on a variety of list serves and web sites over the Internet. The professionals I spoke with for this article – Mark Hutton (former plant breeder for Peto now at University of Maine Extension), C.R. Lawn, Rob Johnston (founder, owner and plant breeder of Johnny's Selected Seed), Frank Morton (Plant breeder and owner, Wild Garden Seed), and Michael Sligh – were in concurrence with the concerns expressed in the online group discussions, first, with regard to the potential decrease in varietal selection for farmers, and second, in the potential acceleration of biotech applications in the vegetable sector.

One can only speculate on Monsanto's motives for purchasing Seminis. We can make educated projections, just as Wall Street financiers have done on news of the acquisition. Financial and agricultural professionals interviewed in the mainstream press, such as Don Basse of the commodity advisory group Agresources, have surmised that the acquisition can be profitable for Monsanto only with the application of biotechnology – as Seminis conventional seed business was nearly half a million dollars in debt and continuing to lose money.

Basse says that it would be logical for Monsanto to use biotech to increase the nutritional value of fruit and vegetables as opposed to focusing on shelf life or devising pest-resistant strains. Monsanto's press release noted that "Biotechnology applications could be an option, and will be evaluated in the context of Monsanto's research-and-development priorities and potential commercial business opportunities." However the main tone of the announcement focused on the trend of nutrition and healthy lifestyles. Monsanto's CEO put it this way, "The addition of Seminis will be an excellent fit for our company as global production of vegetables and fruits, and the trend towards healthier diets, has been growing steadily over the past several years."

"You have to ask yourself why they (Monsanto) would decide to buy this seed company," was the thought first shared by Rob Johnston of Johnny's Selected Seeds, "Their Roundup herbicide patent is expiring, so their future profits are in the biotech traits...I think they're going to push and see if consumers will accept it." C.R. Lawn of Fedco was less certain, feeling that Monsanto would not be bold enough to try and sell such technology to consumers and farmers, particularly after GMO wheat was recently shelved because of the lack of perceived public acceptance. There is also speculation that if Monsanto can slowly start building the GMO vegetable-fruit market, then the debate over GMOs will become a moot point, as they will have made their way onto the plate and thus gaine acceptance (or at least acquiescence).

Even if one does not believe that GMO vegetables will be in the Wendy's salad bar in short order, there is more pressing concern that Seminis will drop many of the hybrid and open-pollinated varieties that regional farmers currently depend upon. Prior to the buyout, the company's main product focus had continued to move towards supplying genetic for the larger centers of production. "It's not like they're still breeding tomatoes for the Northeast" Rob Johnston noted. Still, Johnston conceded that it would be difficult for Johnny's to replace some of the Seminis varieties that their customers turn to year after year, such as Gold Rush Zucchini or King Arthur Pepper. Yet he feels certain that cuts are coming. Johnston was disappointed with the news, in part he said because he likes not only the quality of product but the Seminis breeders themselves, "I worry about the future of their breeding programs, that they (Monsanto) will curtail creative directions and focus them on a Monsanto agenda."

Organic Seed Alliance has received over a dozen e-mails and phone calls from concerned farmers. Minnesota farmer Jim Fruth contacted us for assistance in "dehybridizing" a Seminis hybrid pepper that has recently been dropped. Like many farmers, Fruth has integrated particular varieties into his production and marketing plans and he says he is now without a variety that is "a vital part of my livelihood." Nash Huber of Sequim, Washington, said that, after vast trialing, he had found that Seminis cabbage varieties have excellent post-harvest holding capacity, extending his marketing season and farm profitability. He did not have high hopes of finding replacements.

Mark Hutton worked as a squash and eggplant breeder for Petoseed before it was purchased by Seminis. From his perspective, farmers like Fruth and Huber should start trialing new varieties soon. "Monsanto is going to look at this from a bean-counter perspective. Low margin varieties get dropped, and this means anything that's not for large commercial production."

One seed catalog owner I spoke with believes that farmers should not react to the news by seeking non-Monsanto/Seminis seed sources. He said there is no indication Monsanto will drop these varieties and that rushing to find replacements isn't an answer. "Where are you going to go? Some of these varieties are irreplaceable. Are we really going to drop or boycott some of the best material out there because we don't like Monsanto?" He warned that doing so might only accelerate the downsizing of the Monsanto product list, leaving farmers in a real lurch. "The process of breeding alternatives to these (varieties) is a long, longterm project. And what are you going to plant in the meantime?"

Most of the people I spoke with agreed that there are few options; this is what is making them react to the news so passionately. In a healthy competitive market, a producer has more than one cog to choose from, giving the producer freedom to switch suppliers if they have an issue with their traditional supply chain. In a highly consolidated system, this choice is not easily apparent and may simply not exist.

Consolidations in the seed world are nothing new. The impact is predictable: A few breeders lose their jobs, farmers scramble to find another variety to fit their production system but something eventually comes along, stockholders either make or lose money, and, in the end, food still winds up on the plate of most American households at 7a.m., noon and 6 p.m. We've been here before in recent times, and we've seen even bigger control of seed ownership and distribution (although not in any of our lifetimes).

A-century-and-a-half ago there was only one mega-distributor of seeds in this country. Lobbying and activism brought about its demise. That distributor was the United States government, and the rabble rousers who broke that monopoly were none other than the American Seed Trade Association–whose largest modern financial benefactor is none other than Monsanto.

CONTRACT FARMING IN SEED PRODUCTION

Seed is important in sunflower production. However, quality seeds are not available to meet the growing areas of sunflower in Karnataka. To meet the sunflower seeds, KSSC,KOF, UAS etc. agencies are involved in seed production. The seed production targets are being fulfilled by several ways including contract farming. However, there are several issues raised by different stakeholders about the profitability and feasibility of seed production under contract farming system. Hence, the present study was carried out to address these issues.

Generally, there are three types of contract in agriculture viz."

1. Procurement contracts, under which only sale and purchase conditions are specified
2. Partial contracts, wherein only the contracting firms supply some of the inputs and produce is bought at preagreed prices and
3. Total contracts, under which the contracting firm supplies and manages all the inputs on the farm and farmer is just a supplier of land and labour.

The relevance and importance of each type varies from product to product and these types are not mutually exclusive. Whereas, the first type is generally referred to as marketing contracts, the other two are of production contracts. But, there is a systematic link between product market and factor markets under the contract arrangements as contracts requires a definite quality of produce. Different types of production contracts allocate production and market risks between the producer and the processor in different ways.

The scenario of contract farming can be traced to sugar mills in Karnataka were practicing the contract farming since many decades, where farmers were growing sugarcane at the specified pre-agreed price. In the same manner, corporate sector has introduced it to horticultural and vegetable crops. Similarly, the contract farming of oilseeds for seed production was introduced in Northern Karnataka during 1985-86. It was first introduced in the taluks of Haveri, Ranebennur, Hanagal and Hirekerur of Haveri district by the different firms like KOF, KSSC and private seed companies *viz.*, Mahyco and Sungro. The farmers of this district had engaged since long time in the production of oilseed for seed purpose.

IMPORTANCE OF THE PRESENT STUDY

The KOF in seed production resort to contract farming mainly to have assured supply of genuine seed material in required quantity at the right time, which has been produced under their supervision. On the other hand, the farmers are interested to enter into contract mainly to minimize the price risk, and also to reap higher profits out of this seed production activity over commercial production of crops.

The present investigation is an integrated effort to study all socio-economic aspects and profitability of production of hybrid sunflower seeds in addition to identifying the constraints in its production with an overall view of exploring the possibility of bringing about required improvement. The results obtained from this study would be used to overcome the present limitations in production of hybrid sunflower seeds.

The information on cost and returns structure will guide the producer in readjustment and proper management of resources and to bring down the cost of production at the farm level without affecting the output. The results would help the planners and policy makers in formulating suitable policies for grant of loans and fixation of prices and also throw further light on the avenue for future research in the area of hybrid sunflower seed production.

PRODUCTIVITY OF RESOURCES IN SEED PRODUCTION

Kshirsagar studied the resource use efficiency of different inputs in hybrid cotton seed production using Cobb-Douglas production function. Out of four selected variables in Varalaxmi, the human labour input was significant at one per cent level, while fertilizers and plant protection measures were significant at five per cent level of probability. In the case of H-4 cotton seed, except fertilizer, all selected variables *viz.*, human labour, plant protection measures and irrigation were significant at five per cent level of probability. Patel studied the resource use efficiency of various inputs used in seed production of hybrid cotton in Gujarat. The ratios of MVPs of different factors to respective cost of factors were calculated. It was found that 9.14, 1.23, 5.48, -1.08 and 6.83 ratios of MVP to MFC of bullock labour, human labour, plant protection chemicals, irrigation and other expenditures, respectively.

Pathak studied economics of production and marketing of certified seeds of high yielding varieties of jowar and wheat. He concluded that the factor product relationship, between the output and human labour, bullock labour and manures and fertilizers was positive in jowar seed production. In wheat seed production, it was seen that there was a positive relationship between the output and human labour, manures and fertilizers. Muralidharan studied the resource use efficiency in rice production in Kerala employing the Cobb-Douglas type of production function. Adjusted R2 (0.84) indicated that per cent of the variation in yield of paddy could be explained by the estimated production function. The coefficient of land and human labour were positive and significant at one per cent probability level.

Radha *et al.*evaluated the resource-use efficiency in rice-rice and rice-pulse farming systems in Krishna district of Andhra Pradesh. The results indicated that manures and fertilizers and irrigation were quite productively used in both the farming systems. The sum of production elasticities indicated the operation of constant returns to scale in both farming systems.

Karisomanagoudar employed Cobb-Douglas type of production function in Gadag taluk of Dharwad district to study resource use efficiency in rainfed onion roduction. It was observed that land and labour inputs significantly increased the gross revenue. The seed variable exercised a significant negative influence on earnings from onion. The variables included in the production function explained 96 per cent of the variation in output.

Vasudha assessed the relationship between agricultural output and selected explanatory variables in Karnataka by using Cobb-Douglas production function. It was found that the gross cropped area, fertilizer consumption, bullock labour and human labour were important variables in explaining variation in agriculture output. The study also indicated that the agricultural production was increased at the rate of 3.11 per cent per annum during the period between 1956 and 1983 and fertilizer and irrigation were the predominant contributors for growth.

EVOLUTION OF SEED POLICY IN INDIA

The question arises as to what was the need for such a bill to come into existence. For this it is essential that the history of Indian seed regulation be briefly looked into. Two decades after India's independence, during the 1960s, the formal seed sector in India was dominated by the Public sector. It was in 1961 that the National Seeds Corporation (NSC) was established under the Ministry of Agriculture and was at the centre of seed production of breeders, foundation and certified seeds and their quality control. In furtherance of their control in the seeds sector, the National Seeds Project (NSP) was undertaken by the Indian Government in 1967 along with the assistance of the World Bank.

While the NSP did set up huge seed processing plants in order to provide 'certified' seeds of food crops, mainly self-pollinating to farmers, most of these plants operated well below capacity and were stated to be examples of 'faulty technology being pushed into India'. This was, of course, not the only big public sector project that failed to achieve the desired goals and gradually a distrust of the public sector set in. However, the private sector has not proved much more efficient or beneficial either for the cultivator or for the consumer. This led to the National Seed Policy of 1988, which involved a US $ 150m loan from the World Bank to help privatize the Indian Seed Industry. At this time, the import of seeds was still restricted but this sector was gradually opened up, to allow more private participation. Further, after India signed the GATT agreement and joined the WTO, these agreements required that India make some changes to its law, especially regarding

INTELLECTUAL PROPERTY RIGHTS

These requirements were met through the Protection for Plant. In 2002, a new National Seed Policy was released, and as mentioned above, to meet the goals of this policy the new Seed Bill was drafted and tabled in Parliament in 2004.

The objective of the new seed policy seems to be to reduce the direct involvement of government in seed production and marketing, and to actively encourage the private sector to engage in research and development of new varieties. The dominance of the public sector is blamed for the backwardness of Indian agriculture, and one of the stated aims of the National Seed Policy, 2002 was to encourage more private participation in agriculture and seed production, specifically, to complement the existing structures and toreplace them, when necessary. However, this is not to say that the Seed Policy is an insidious conspiracy to betray small and medium-scale farmers. Liberalization has been targeted towards certain components of national seed policies, retaining regulation of some components to safeguard national interests. As the Food and Adulteration

Organisation has noted, the change in the structure from one geared towards the public sector to a more private sector – centric approach is an extremely complex one and would take time. Public and private sectors need to complement each other, perhaps on the basis of a division between cash crops and essential food crops. It is a fact that neither the private nor the public sector can fulfil India's agricultural requirements by itself. Only effective cooperation and coordination will allow farmers to have access to quality seed and thus contribute to sustainable agriculture and food security. The National Seed Policy, 2002 clearly identifies the twin aims of encouraging the seed industry, especially the domestic industry and of ensuring maximum prosperity and security for farmers. A number of the National Seed Policy's recommendations have been addressed in the Protection of Plant Varieties and Farmers' Rights Act, 2002, including the establishment of a National Gene Fund and a Plant Varieties' authority to regulate the quality of seeds in the country. The further aims of the National Seed Policy that include building up infrastructure, ensuring good quality of seeds and facilitating international trade in seeds, are sought to be addressed through the proposed Seeds Bill.

THE SEED BILL

At the very outset, the preamble of the bill makes clear its intention. The Bill states that it is to "provide for regulating the quality of seeds for sale, import and export and to facilitate production and supply of seeds of quality and for matters connected therewith or incidental thereto".

RATIONALE FOR THE SEED BILL

The new seed bill is a tool to address the grievances and the concerns which the Seed Bill of 1966 does not cover. Even though a large majority of our population depends on agriculture for their livelihood, agriculture in India remains, to a large extent, backward and relatively unproductive. The much touted Green Revolution did have a positive impact but the lot of marginal and small farmers has not improved much over the past few decades. As such, a need was felt for using new techniques and methods to increase the productivity of Indian agriculture. At the same time the Bio-Technology sector came up with promises of extremely productive Genetically Modified (GM) Crops.

These new scientifically manipulated crops caught the imagination of the Indian Government, and to some extent, that of farmers as well. On one hand, there were reports of farmers' suicides owing to failure of crops, including Bt Cotton, and on the other, stories of farmers who found the highway to prosperity through the use of the same Bt Cotton. It is believed that this new technology has the potential to improve living standards, and as such, several powerful groups support the commercialization of such GM crops (especially Bt cotton), including the Indian Council of Agricultural Research, Ministry of Environment and Forests,

About 50% of the total investment in agricultural research in developed countries is contributed by private sector. The idea behind the Seeds Bill, 2004 is that effective implementation of this new law can be expected to promote private plant breeding in the country in the long run. The major rationale behind the policy is the hope that these developments would provide Indian farmers multiple choices and increased access to improved seeds. As such, the Seed Bill 2004 also seeks to address the concerns of the Seed Industry. The Seed Association of India and the Association of Seed Industries raised certain demands at the National Seeds seminar organized by them in 2005. They demanded a level playing field for the private sector, for subsidies and support to the private sector for R&D (specifically to facilitate exports). Another major demand was that seeds be taken out from the purview of the Consumer Protection Act, 1986 and that a scientific system of scrutinizing claims, along with a system of crop insurance, should be developed to study the causes of crop failure.

While the Seeds Bill, 2004 does take care of the infrastructural demands but it retained the right of the farmer to go to the Consumer Courts under the Consumer Protection Act, 1986. One cannot be sure if, in the present situation in India, where a majority of farmers are not only small landholders, but they are also illiterate and/or uneducated, a system of crop insurance can work to their advantage. It was in this atmosphere that the National Seed Policy was formulated in 2002 with the National Seed Bill being drafted.

SALE OF SEEDS

The Seed Bill aims to regulate the quality of seeds sold in India. It requires registration of all varieties of seeds sold in India. The Bill states in S.1 (3) that it applies to every dealer and every producer of seed except when the seed is produced by him for his own use and not for sale. Thus, the only exception to the rule is the exemption granted to farmers to use and sell seeds from their own farms, as long as such seeds are sold unbranded. However, such seeds will also have to meet the minimum standards set for registered seeds, a requirement which will obviously be hard to fulfill for a small farmer and probably as hard to detect for the enforcement authorities. This has been criticized as an assault on the freedom of farmers, a denial of their time-honoured rights, as an undermining of seed sovereignty of farming communities. Farmers care for their own seed quality more than a centralized authority can.

Regulation of farmers' own seed varieties needs to be left to farmers. That is why we have established Community Biodiversity Registers and Jaiv Panchayats. Further, in India, where up to 66% of the seed requirements are met by farmers, declaring all of this to be non–commercial does not make much sense. Farmers' Rights have to be recognized for one major reason; they are the original source of the germ-plasm. Under most IPR Regimes all over the world, farmers are not recognized as innovators, and it is the providers of technology who acquire Intellectual

Property Rights. For the sake of conservation and to give the farmers a fair share of the economic benefits arising out of usage of their seeds, there should be provisions for recognizing and awarding community rights. The Seeds Bill, 2004 has no such provision although the Plant Variety Protection and Farmers' Rights (PVPFR Act) does include provisions for compensation for communities for their role in research and development.

A feature of the policy and the bill is that they start off with complete acceptance of GM crops. There is no longer space for argumentation and debate on that particular issue. India, as mentioned above, has taken to GM crops in a big way. GM crops have not found the same level of acceptance everywhere. Several EU countries continue to block entry of GM products based on concerns for their environment and public health. The seed policies of the European Union have been described as 'control-oriented' seed legislations. Seed quality control and certification agencies in Europe are run by the Government in countries like Germany or by an independent foundation like in the Netherland. These independent foundations are usually a representative of farmers' societies, and owing to the sheer magnitude of India's population, may not be the ideal solution for Indian agriculture.

Also, in Europe, especially in the horticultural sector, certification agencies can perform a full control or they may concentrate on certifying the internal quality control operations of the seed companies, rather than certifying each and every seed lot throughout the seed production-marketing-user chain. All these systems monitor GM crops and their effects from the stage of production of seeds till food processing and consumption.

Once GM crops are allowed entry into the market in a major way it becomes necessary to protect the Intellectual Property Rights (IPRs) of the organizations and industries which have patented such seeds. The Plant Variety Protection and Farmers' Rights (PVPFR) Act was passed in 2001 to take care of this need. The government has provided sufficient protection to the Intellectual Property Rights of the seed developers. It is possible that the implementation of the IPR regime will increase private sector activity. This would invariably lead to an increase in prices, which would adversely affect resource-poor farmers in marginal areas. Thus, the government will have to monitor the seed sector very closely and should effectively intervene if the market fails to serve the farmers. This requires more decentralization and flexibility in operations of public seed agencies.

Although, registration and certification of seeds under the Seeds Bill do raise certain questions regarding Intellectual Property Rights, the issues regarding Intellectual Property Rights are dealt with by the PVPFR Act. Through certification and registration, the Seed Bill only safeguards such rights acquired under the PVPFR Act. Under both Statutes, a degree of leeway is given to farmers. Under the Seed Bill, farmers are free to use and sell their own unbranded seeds while under the PVPFR Act, under Section 42, farmers have been given protection in cases of 'innocent infringement', wherein if the farmer is not aware of breeder rights, at the time of infringement, she will not be liable for infringement. It is perhaps a testament to the lack of progress on the PVPFR Act, that a need is still felt for a comprehensive database of seeds in India. The PVPFR Act itself has allowed for a National Register of Plant Varieties to be set up. Now, the proposed Seed Bill incorporates the requirement for a National Register of Seeds. The Bill proposes to establish a Central Seed Committee (CSC), which is roughly on the lines of the National Seeds Board envisaged by the Seed Policy, 2002. The CSC will appoint a Registration Sub-Committee, which shall maintain the National Register of Seeds. Further, State Seed Certification Agencies will be set up in each State. These Agencies will have the authority to certify seeds. Further, certain accredited agencies may also be allowed self-certification.

This is a risky proposition at best. Neither the public nor the private sector enjoy an unimpeachable reputation in India. It would be an unsurprising development should wake one up to newspaper headlines of companies and agencies having fudged data to grant certification to certain varieties. The State needs to be the ultimate overseer and guardian of farmers' rights and public health. In the United States, for example, seeds and the crops that they grow into have to secure approval from one, two or three government agencies, as the case may be, before being marketed. The Food and Drug Administration has to review whether the food that will ultimately be produce will be safe to eat, the U.S.

Department of Agriculture's Animal and Plant Health Inspection Service reviews whether the crop is safe to grow, and the Environmentl Protection Agency studies whether the new crop is safe for the environment. Another aim of the National Seed Policy was to streamline the monitoring mechanisms regarding seeds and crops. As such, under the Seed Bill any type of seed for sale will have to be registered with the Registration Sub-Committee. The registration will be valid for 15 years for annual/biennial crops and 18 years for long duration perennial crops. On expiry, the registration can be renewed for a similar period. This is a problematic proposition as usually IPRs are granted for a specific period, whereas in this case, the Registration sub–committee can renew the registration ad infinitum. All registered seeds will need to meet minimum standards with respect to the proportion of seeds that must germinate, levels of genetic and physical purity, and the permitted proportion of diseased seeds.

No transgenic variety of seed can be registered under the proposed Bill unless the applicant has obtained clearance under the provisions of the Environment (Protection) Act, 1986. Any variety of seed that contains any technology considered harmful or potentially harmful cannot be registered in India under the proposed scheme. It is hoped that this terminology excludes the use of terminator seeds in India, as such seeds cause great harm, especially to small farmers. Also, a major lacuna under the Bill is that there is requirement for disclosure of 'parental lines' at the time of registration. Such a disclosure is essential for the sake of information and safety. Under the proposed Bill, it is the responsibility of the Central Seeds Committee to set minimum standards for seeds and to decide which seeds are harmful to the environment or dangerous to public health. Further, Central and State Seed Testing Laboratories will be set up. These Laboratories will analyse all varieties of seeds. Lastly, the respective State Governments will need to appoint Seed Inspectors. The Seed Inspectors would be given powers of search and seizure relating to an offence under the proposed Bill. This is quite a worrying component. Under these provisions, Seed Inspectors have been given wide powers and unlike under the Code of Criminal Procedure, no warrant is necessary nor are procedural safeguards applicable. It is not necessary to outline how this particular provision develops scope for abuse of power, but the safeguards against harassment by functionaries of the State exist for a very good reason, and such safeguards should be included here as well.

REMEDIES AVAILABLE TO FARMERS

Apart from the success stories publicized by the Bio-tech industry to illustrate the success of their seeds, there are numerous stories of failed crops and economic ruin as well. As such, one of the most important provisions in the Bill as far as Farmers' Rights are concerned is regarding the remedy available to farmers in case of defective seeds. Under the Seed Bill, in case the seed supplied to the farmer is found to be of inferior quality and does not give the desired results, the farmer can approach the Consumer Courts under the Consumer Protection Act, for compensation and redressal. Also, under the PVPFR Act, under Section 39 (2) of that Act, a farmer or a farmer's organization can claim compensation if a variety fails to give the expected performance under given conditions. If the performance of the variety is not found to be as claimed by the breeder, the Authority can deal with claims of failure of performance and could decide about such claims independently, instead of the courts. However, it is not clear whether the quantum of compensation awarded would cover only the cost of the seed but that of the crop as well.

WIDESPREAD OPPOSITION TO THE SEED BILL

The Seed Bill, 2004 has met with opposition from farmers' organizations, political parties as well as large swathes of society. The All India Kisan Sabha (AIKS) has come out openly against this Bill.

The AIKS has decided to organise widespread campaigns and mass actions at village, mandal and district levels throughout the country in May and June on the issues concerning peasants in particular and people in general. Political parties have also come out against this Bill. The Communist Parties have denounced this bill as being anti-farmer. On the other end of the political spectrum, B.S. Yediyurappa, Deputy Chief Minister of Karnataka and a member of the Bharatiya Janta Party (BJP), has also condemned the new Seed Bill. Bharat Krishak Samaj (BKS), the ruling Congress party's farmers' outfit, has also opposed the legislation. Apart from concerns expressed by prominent activists such as Vandana Shiva and Devinder Sharma, the Bill has also been blasted by mainstream newspapers. Dr. Suman Sahai, a noted geneticist and President of the Gene Campaign, has pointed out that the PVPFR Act, drafted to comply with India's obligations under the TRIPS Agreement, does a much better job of regulation and protection of farmer's rights.

She notes, "Key differences between the Seed Bill and the PVPFR relate to declaring the origins (parentage) of the variety, the conditions for multi-location testing and who will conduct these tests, the level of transparency maintained on grant of registration, price control and the treatment of farmer varieties. While the PVPFR requires the declaration of the origin of the variety with pedigree details, the Seed Bill does not. The Seed Bill does not grant any recognition to the contribution of farmers as conservers of agricultural gene pools, or as breeder of successful new varieties This violates the provisions of the Convention on Biological Diversity and the PVPFR according to which if farmer varieties have been used in breeding a new variety, farmers are entitled to share in the profits through the benefit sharing mechanism.

The industry gets its way with seed pricing in the Seed Bill; there is no mechanism to regulate seed supply or seed price unlike the PVPFR which has a clear provision for compulsory licensing. Compulsory licensing safeguards the interests of the farming community since it places the responsibility of ensuring an adequate seed supply at reasonable price on the government. As a result of the Seed Bill, we could have high cost of seeds fixed arbitrarily by the seed companies, leaving the government with no means to control the price. It could also mean that seed providers are under no obligation to ensure a reasonable seed supply to farmers." This is the most damning indictment of the proposed Seed Bill.

The fact is that a legislation that adequately protects farmers' rights as well as the Intellectual Property Rights of the Industry is already in place. The National Gene Fund and the recognition of Community Rights under S. 41(1) of that Act are path breaking developments and go a log way in convincing skeptics of the fairness of the National Seeds Policy. Now, after hectic lobbying by the Seed Industry the new Seed Bill actually seeks to dilute the PVPFR Act. The PVPFR Act was drafted to ensure compliance with Article 27.3 (b) of the TRIPS agreement. After that agreement, India's obligations to the WTO as regard intellectual property rights of seed companies have ended. There is no need to change the law once again. The government needs to take another look at the proposed bill to ensure that there is no overlap with the PVPFR Act, and especially that the beneficial provisions of that Act are not diluted.

8

Seed Treatment

Most seed treatment products are fungicides or insecticides applied to seed before planting. Fungicides are used to control diseases of seeds and seedlings; insecticides are used to control insect pests. Some seed treatment products are sold as combinations of fungicide and insecticide.

Fungicidal seed treatments are used for three reasons:

- To control soil-borne fungal disease organisms that cause seed rots, damping-off, seedling blights and root rot;
- To control fungal pathogens that are surface-borne on the seed, such as those that cause covered smuts of barley and oats, bunt of wheat, black point of cereal grains, and seed-borne safflower rust;
- To control internally seed-borne fungal pathogens such as the loose smut fungi of cereals.

Most fungicidal seed treatments do not control bacterial pathogens and most will not control all types of fungal diseases, so it is important to carefully choose the treatment that provides the best control of the disease organisms present on the seed or potentially present in the soil.

The degree of control will vary with product, rate, environmental conditions and disease organisms present. Some systemic fungicidal seed treatments may also provide protection against early-season infection by leaf diseases. Fungicide-insecticide combination products or an addition of insecticide for wireworm control should be considered if planting newly opened land or land that has had a history of wireworms. Consult current recommendations for insecticides registered for wireworm control. A fungicide-insecticide combination also may be useful for dry beans. The insecticide used on dry beans should be one that provides control of the seed corn maggot.

TECHNOLOGY OF SEED TREATMENT

Successful seed treatment is a complex story as it depends on a number of interacting factors. Full compliance with all of them is essential to achieve a high quality seed treatment. It all starts with the seed being treated needing to be clean and of high quality itself. The seed treatment formulation has to be stable and have basic adhesive properties; the film coating has to complete the adhesiveness in order to ensure even distribution of the seed treatment product on the seeds and to provide good seed flowability. The slurry recipe has to be adapted to the seeds, and the volume needs to be adjusted to achieve good coverage and a homogeneous distribution among the seeds. Finally, the equipment must be able to be adjusted precisely and has to work reliably. The treatment will be of high quality only if the operator applies exactly the right dose of both seed treatment product and film coating.

Seed treatment technology has come a long way since the use of salt brine in the mid-1600s. Today, seed treatments deliver clear environmental, economic and social benefits, making the technology a perfect tool for sustainable agriculture. With a growing world population, now more than ever it's critical that farmers have access to the tools that will help them grow more food while protecting the environment. And seed treatments are one of these tools.

"By utilizing modern agriculture technologies, farmers will be able to boost yields, conserve water usage and protect biodiversity," says Keith Jones, director of stewardship and sustainable agriculture for CropLife International based in Brussels, Belgium. "Seed treatments represent one tool on which many sustainable agriculture technologies rely upon. By protecting seeds from planting to emergence, seed treatments can improve stand establishment and increase potential yield."

Helmut Schramm, head of the seed treatment business at Bayer CropScience in Monheim, Germany, agrees, saying seed treatments deliver clear environmental, economic and social benefits, making this technology a perfect tool for sustainable agriculture.

"Innovative seed treatment technology represents an environmentally-sound approach to crop protection," he says. "Treating the seed provides a targeted and effective means of application that helps increase yields, safeguard our environment and ensure a sustainable means of crop production." Over the years, seed treatments have evolved from simply protecting the seed to helping improve plant stand and early plant health.

"Seed treatments are increasingly designed to also enhance plant emergence, growth or nutrition efficiency, which, along with crop protection, lead to a more vigorous and uniform crop," says Schramm. "This forms the optimal base for a high-quality crop which can fully exploit its yield potential."

Greg Lamka, chair of the International Seed Federation Seed Treatment and Environment Committee, says that in order to maximize yield, it's important to have a full and uniform stand. The STEC committee was established in the 1990s to raise the seed industry's awareness about the use of different seed treatments and to promote a better understanding of how production could be improved and made more efficient. The committee consists of seed companies and crops protection companies who wish to promote the safe and effective use of seed treatment products. "Growers are paying more for the seed, so expectations are rising about the performance of our products. Seed treatments are a way to ensure that the products will perform to their maximum, based on the environment that they're put into," says Lamka, adding the vast majority of the seed used in developed countries is treated.

COMBINATION OF INGREDIENTS

And what do seed treatment products consist of? The most important thing, of course, is the active ingredient which is responsible for the fungicidal and/or insecticidal effect. Yet a seed treatment product contains many other components too, the most important of which are adhesive substances (to make the active ingredient stick to surface of the seed), dispersion substances (to allow for even distribution of the active ingredient), and colorants (so that it can be seen immediately whether or not even distribution of the seed dressing has been achieved). These and other important properties such as seed flowability are supported and complemented by the co-application of a proper film coating designed for a specific seed treatment product.

MODERN TECHNIQUES

Nowadays, modern seed treatment machines have been designed to a high standard for treatment of large quantities of seed . However, different seed treatment formulations and different seeds sometimes require different machines. The most common seed treatment formulations worldwide are flowable concentrates, wettable powders and liquids. Advanced seed dressing methods include film coating and pelleting. Bayer CropScience's most recent development is water-borne coatings, seed treatments in the form of a water-based suspension which are particularly user-friendly. They generate neither dust nor solvent vapors during use, and machines and equipment are easy to clean with just water.

SEED TREATMENT APPLICATION

Fungicide seed treatment products come in a variety of formulations and in a variety of packaging sizes and types. Some are registered for use only by commercial applicators using closed application systems, others are readily available for on-farm use as dusts, slurries, water soluble bags, or liquid ready-to-use-formulations. Whatever the formulation used or application method chosen, some precautions should be taken to assure applicator safety and appropriate seed coverage.

CAUTIONS

Follow label directions when handling seed treatment chemicals. These products are potentially poisonous if mishandled or misused. Extreme caution must be used when handling seed treatment chemicals: some are toxic, others may be irritating. An approved chemical respirator and goggles are recommended even if not specifically required by the fungicide label.

The rate of application prescribed by the label must be used: over treatment may injure the seed and under treatment may not provide good disease control. To apply the correct rate, it is essential to calibrate application equipment carefully and to check calibration frequently. Metering cups of commercial applicators should be cleaned daily to prevent a buildup of chemical that might result in reduced application rates. An auger which has been used to treat seed cannot be cleaned up sufficiently for use in augering grain for food or feed. Once an auger is used for seed treatment, it should be used only for treatment or augering seed for planting. It should not be used to auger grain used for food or feed. Treated seed should not be used for food or feed, and treated grain should not contaminate grain delivered to elevators or be placed in bins or in trucks delivering to elevators. Containers should be triple rinsed with the rinse water added to the treatment mixture. The rinsed containers should be punctured and crushed for disposal in an approved landfill.

APPLICATION METHODS

SEED DRESSING

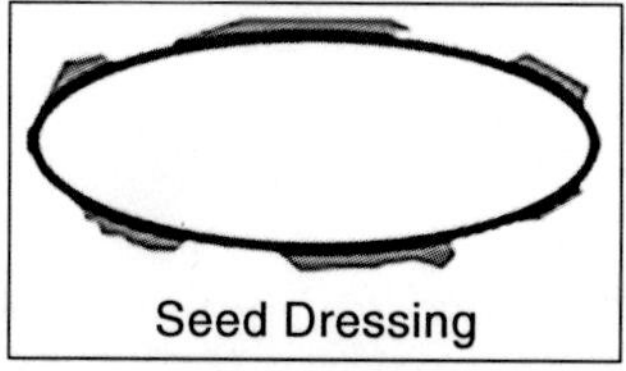
Seed Dressing

Application using simple application machinery,

FILM COATING

Film coatings are applied together with the insecticidal and/or fungicidal seed treatment product to improve the application quality. Proper film coatings applied at the correct use rate reduce abrasion, dust formation and hence the loss of active ingredient during application, packaging and sowing. They improve the even distribution of the seed treatment products on the seeds and restore good seed flow ability/sowability.

Film coatings are part of professional seed treatment for field crops and vegetables, and are applied mainly by professional seed companies. In order to confer all the technical properties such as reduced abrasion and dust formation, good flowability, and improved seed coverage and coloration, modern film coatings are quite complex and sophisticated products which contain polymers, loading materials and wetting and stability agents in combination with pigments and shine agents.

The combination of these components and the exact recipe strongly depend on the detailed requirements, the crop and the seed treatment product used.

In addition, film coatings must not hinder water uptake by the seeds, their germination or field emergence. Film coatings form a very thin film and do not change the size, shape or weight of the seeds.

PELLETING

Pelleting has two main purposes. One is to give seeds with an uneven surface a uniform and homogeneous size and shape, *e.g.* for sugar beets and fodder beets. The second purpose of pelleting is to increase the size and/or weight of very small seeds such as vegetable or grass seeds. In both cases the intention is to adapt and change the shape, size and/or weight of the seeds to allow precision sowing with modern equipment.

The inert materials used for the pelleting process must be capable of forming a robust and stable pellet but at the same time must not hinder water uptake by the seeds and hence germination and emergence. Seed treatment products are usually applied to pelleted seeds in a second step and always in combination with a film coating.

PELLETING AND COATING

The term describes the sequential application of different film coatings in combination with different seed treatment products. It can be used with either pelleted or non-pelleted seeds.

MULTILAYER COATING

A highly sophisticated method allowing sequential application of multilayer materials, including the incorporation of fungicides and insecticides.

GIVING SEEDS A FILM COATING

Film coating products contain polymers, dyes and surfactants and form a fine air- and water-permeable film that improves the distribution and retention of crop protection agents on the seed surface. The coating products contribute to a significant reduction of the amount of dust released during the application of the crop protection agents to the seed and during handling and use of the treated seed on the farm. This means the operator is better protected.The film coating also improves the flowability of the seed, *i.e.* the ability of seed to flow instead of clumping in the treatment station or sower, which means easier production on an industrial scale. Film coating also allows sowing rates to be managed accurately. The coloration of the coating gives the seed an attractive, glossy appearance, but also has the practical use of allowing differentiation between varieties and between different types of treatment. Film coatings do not significantly modify the shape and weight of the seed.

PELLETING FOR UNIFORM SIZES AND SHAPE

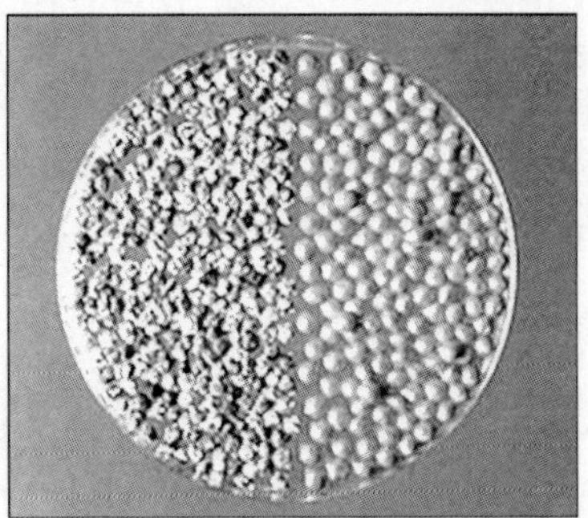

Surrounding a seed with a thick coat, the pelleting process is designed to make seeds uniform in size, shape and weight so that they can be precision-sown with modern drilling machinery. The other advantage of pelleting is the additional protection it provides to the seed. The pelleting material must be robust enough to remain in place during handling, transport and sowing. Pelleting guarantees that the active ingredients are distributed evenly among individual seeds within a seed lot. The application of a covering layer of fungicide and/or insecticide ensures that exactly the right dose is available at the right time.

ADVANTAGES

Seed treatments refer to the application of crop protection products directly to a seed to protect it from seed-borne and soil-borne pathogens and insects. The time between planting and emergence is the most vulnerable stage of plant development. Once in the soil, the seed is susceptible to damage from insects, pests, bacterial pathogens and fungi. Seed treatments allow farmers to protect against these threats during planting, says Jones.

When possible, farmers plant earlier than they used to in order to maximize yields, and often reduce tillage or decide not to till at all, says Lamka, noting that both of these practices significantly impact the seed bed.

"The earlier you plant, the more often it's going to be cold and wet. The more plant debris you have laying on the surface of the soil, the colder and wetter the seed bed will be and the more plant disease inoculums that will be present. The colder and wetter the seed bed, the slower the seed will germinate, and the slower it comes up," he says, adding that this provides more opportunity for fungi to attack and kill seed, or greatly reduce its health. This risk can be decreased by using seed treatments.

Seed treatments are also a highly targeted way of applying pesticides, notes Schramm. "Instead of spraying the entire field area, less than one per cent is treated, and so only insects and pathogens that forage on the plants are exposed," he says. "Therefore, beneficial species and other species that live on and around the plants are protected." Less product use per area also leads to decreased risk of off-crop drift, which consequently has a reduced impact on species in adjacent areas, he adds.

Lamka agrees, saying that in the past, if you had a problem with insects or disease, you would make a foliar or granular application. However, when you use seed treatment, you end up using less product and you're burying the product underground. "So we've greatly reduced our impact to the environment by using these very small amounts of focused material as a seed treatment," he says.

Many of the new chemistries used for treatments are systemic, explains Lamka. They come off the seed coat into the soil, and are absorbed by the seedling through the root system as it grows. These products often protect the seed and the seedling for approximately three weeks after emergence.

Farmers prefer seed treatment over crop spraying because it is more effective in terms of crop protection, and generates vigorous plants and increasing yields, while being more cost-efficient, says Schramm.

"Entire field spraying can be spared, reducing the use of fossil fuels (and the greenhouse gas emissions associated with their use) as some foliar sprays are no longer necessary," he says. "So this addresses the economic pillars of sustainability, while complementing the technology's environmental benefits." Seed treatments are an environmentally safe way to protect plants because of the small use of active ingredients per unit of land area, says Lamka, who is also the global senior manager of seed applied technologies for Pioneer Hi-Bred in Johnston, Iowa.

"The products are more environmentally safe than they've ever been in history. Using seed treatments is a good stewardship practice," he says.

Seed treatments go back thousands of years, notes Lamka, to when they put salt brine on wheat seed to get rid of certain seed-borne diseases. Years ago, mercury was also used as a seed treatment because it was very effective at killing insects that were attacking the seed, but it's a toxic product to all living organisms. These early toxic products have been banned and taken off the market. "Today we're using much safer products. They are safer for the people handling them, safer for the environment and much safer for the seed itself," says Lamka.

CEREAL SEED TREATMENTS

Fungicidal seed treatments help control soil-borne pathogens that cause seed decay, seedling blight and root rot. Control of these diseases may result in better stands, more vigorous seedlings, and increased yields. Protectant fungicides such as captan, maneb, PCNB, thiram, or fludioxonil (Maxim) help control most types of soil-borne pathogens, except for common root rot and take-all. Protectant fungicides containing captan, maneb, or thiram are sold under various trade names. Fungicidal seed treatments to protect against the soil-borne fungi that cause common root rot and take all will be discussed under barley seed treatment.

Fungicidal seed treatment controls most, but not all, of the seed-borne diseases of small grains. Seed treatment should be used when the seed is contaminated with smut, scab, or black point fungi. Specific recommendations for these diseases are discussed under barley and wheat. No seed treatment is a substitute for good seed.

Barley

Barley has three smuts: covered smut, black semi-loose smut (nigra smut), and loose smut. Covered smut and black semi-loose smut are surface-borne fungal pathogens that infect the emerging seedling. These two smuts can be controlled by various protectant fungicides. Loose smut infects the embryo of the seed before harvest.

Protectant fungicides do not control loose smut—only the systemic fungicides carboxin (Vitavax or Enhance), triadimenol (Baytan) or tebuconazole (Raxil) will control loose smut in barley. These same products also will control the covered and semi-loose smuts of barley.

Covered Smut

Seed of barley varieties susceptible to covered smut should be treated with a protectant fungicide or with carboxin, triadimenol or tebuconazole.

At present, all barley varieties recommended for production in North Dakota must be considered susceptible.

Black Semi-loose Smut

Many varieties are susceptible to this smut. It is indistinguishable from loose smut in the field, but it is borne on the seed surface and infects seedlings just as covered smut does. It is not detected by the embryo test, because it does not infect the embryo. It is controlled by seed treatment with a protectant fungicide or with carboxin, triadimenol, or tebuconazole.

Loose Smut

Many of the barley varieties released in the late 1970s and early 1980s were resistant to the races of the loose smut fungus present at that time. In the mid-1980s, a new race of the loose smut fungus was detected in North Dakota. Seed treatment for loose smut in barley planted for seed production is especially important. Seed samples of susceptible barley varieties should be sent to the North Dakota State Seed Department for an embryo test to determine the percentage of loose smut infection. Losses from loose smut are about equal to the per cent infection; 5 per cent loose smut represents a yield loss of about 4.3 per cent. If the embryo test shows 2 per cent or greater infection, the seed should be treated with carboxin, triadimenol or tebuconazole, or loose smut-free seed should be used for planting.

Common Root Rot

The fungus that causes common root rot is soil-borne and is widespread in North Dakota soils. The common root rot fungus increases under barley or wheat culture, and it often causes mature plant root rot. The fungus also may cause severe root infection in the seedling stage and if splashed to the leaves and head causes spot blotch and black-point infections, respectively.

A number of seed treatment fungicides are now labeled for wheat and barley for suppression of mature plant root rot and also for suppression of seedling blight due to the common root rot fungus. Imazalil is available under several trade names. Triadimenol (Baytan) also is registered for suppression of seedling blight due to common root rot. Tebuconazole (Raxil) is registered for wheat and barley, and difenoconazole (Dividend) is registered only for wheat at the time of publication of this circular.

The conditions in which seed treatment for common root rot would be most beneficial are: where continuous wheat or barley is grown, or if short rotations between these susceptible crops are practiced, and in soils where moisture stress is very likely.

Barley Stripe

This fungus disease is rarely seen in North Dakota but occasionally appears in two-row barley if contaminated seed is planted. The disease causes distinct yellow, then brown stripes that run the entire length of the leaves. The stripes join together and the leaves become split and frayed or shredded in appearance. The fungus is seed-borne in the hull and seed coat. If seed from a suspected disease source is used, seed treatment with a carboxin + thiram or tebuconazole product provides some control of the barley stripe fungus. The systemic fungicide imazalil provides good control of barley stripe.

CORN

Corn seed is especially susceptible to attack by soil-borne pathogens when sown in cold wet soil, when the seed is in poor condition, when it is mechanically injured, or if it has been stored for two years or more.

Seed treatment will protect against seed rot and reduce the danger of seedling blight. Sweet corn is more susceptible to attack than field corn, but both should be treated. Most field corn is already treated when purchased. A number of protectant and systemic fungicides are registered for control of seedling blights and seed rot in corn.

WHEAT

Bunt or Stinking Smut

The fungus causing this disease adheres to the seed surface and then infects the emerging wheat seedling. The bunt fungus has a fishy odor and imparts the same odor to flour made from bunted kernels. The price of bunted wheat is discounted for this reason. Bunt is not known to occur in North Dakota at the present time, nor has it occurred in recent years. However, it occurs in other wheat-producing states in the Great Plains.

Growers who purchase seed from out of state should use a seed treatment fungicide, since out-of-state seed could be infested with (carrying) bunt. Growers who have their crop custom combined and then save their own seed should use a fungicide, since the combine could be contaminated with bunt spores or bunt balls from states to the south. Many protectant, systemic, or combination products are available to control bunt.

Loose Smut

Loose smut of wheat infects the embryo, as with loose smut of barley. However, the wheat loose smut pathogen does not infect barley and the barley loose smut pathogen does not infect wheat. No embryo test is available for detection of loose smut in wheat. Counts of loose smut-infected heads can be made in wheat and durum fields to estimate the per cent infection in the crop. Counts should be made at flowering time, as loose smut heads are hard to detect later. The counts provide information on the per cent infection in the seed that was planted but are not a reliable estimate of the amount of infection in the seed of the harvested crop. If weather conditions are favorable, *i.e.* cool and wet, infection in the harvested seed could be considerably higher than in the seed that was planted. Carboxin, difenoconazole, tebuconazole, and triadimenol seed treatments control wheat loose smut.

Scab and Black-point

Scab and black-point are fungal diseases that attack both wheat and barley seed. The scab fungus also may infect oats. The scab fungus infects the kernels during flowering if warm, wet conditions prevail. Black-point infection occurs from heading to maturity and also is favored by warm, wet weather.

Scabby seed is shriveled, light in test weight, and often has a chalky white or pink discolouration. When planted, scabby seed has poor germination and poor vigour. The scab fungus does not grow systemically from the seed through the plant to cause subsequent head scab infection. The source of head infection is spores from infected small grain or corn residue. Black-pointed seed has a black to brown discolouration of the embryo or germ end, a discolouration which may extend around the kernel into the crease. When planted, black-pointed seed also may have low germination; if the seed germinates, the seedling roots are often infected by the fungus. Protectant seed treatments or protectants in combination with systemics, or systemic products alone have provided significant improvement in stand and vigour. Yield increases also have occurred with treatment of scabby and black-pointed grain. Seed treatment helps assure good stands and seedling vigour; it will not prevent possible scab or black-point infections of the head during the growing season.

Take-all

Take-all is a root disease of barley and wheat caused by a fungus that thrives in very wet soils. Take-all generally is not a serious or common disease in barley or wheat in North Dakota, unless the crop is grown under irrigation. Triadimenol (Baytan), difenoconazole (Dividend) and tebuconazole (Raxil) are registered for suppression of this disease.

PYTHIUM

Damage to wheat from the *Pythium* fungus has not been well documented in North Dakota, but the fungus is common in agricultural soils and generally does more damage where wheat is planted into crop residues and soil temperatures are cool and soil moisture is high. Several seed treatment fungicides are registered for control of *Pythium* spp. in wheat, including difenoconazole + mefenoxam (Dividend XL), mefenoxam (Apron XL), and metalaxyl (Allegiance).

OATS

Smuts on oats have not been reported as a recent problem in North

Dakota, although they have been serious in other states in the region. Oat smut can cause severe losses. Distinguishing between covered and loose smuts of oats in the field is not easy. Oat loose smut is more difficult to control with protectant fungicides than oat covered smut, because the loose smut spores are lodged under the hulls where they are difficult to reach. Carboxin and tebuconazole are registered for oat loose smut control.

LEGUMES

Alfalfa and Small Seeded Legumes

Alfalfa, clover and other small seeded legumes may be treated for seedling blight diseases and damping off. The extent of occurrence of these disease problems in North Dakota has not been documented. Captan or thiram products are registered for seedling blight control, while mefonaxam, metalaxyl, and oxadixyl products are registered for control of *Pythium* damping off and early season infections by *Phythophthora.*

Soybeans

Generally, seed treatment is not required for soybeans. However, it may pay to treat seed if: the germination is below 85 per cent; the seeds are contaminated with fungi; the seeds are badly weathered or injured; or the seed coats are broken. If captan or PCNB treated seeds are to be inoculated with *Rhizobium* bacteria, the inoculant should be applied in-furrow.

Thiram, fludioxonil, metalaxyl and mefenoxam have little or no effect on *Rhizobium* bacteria. Generally, inoculant is not used on soybeans planted on land that was previously cropped to soybeans.

In fields with *Rhizoctonia* problems, a seed treatment containing carboxin, chloroneb, or PCNB should be used. Seed infested with Sclerotinia should be treated with fludioxonil, which will control infection of seedlings by the Sclerotinia fungus.

Dry Edible Beans

Most dry bean seed is treated prior to sale, often with a combination of fungicide, insecticide and bactericide. A fungicide protects against seed- and soil-borne fungi. The bactericide streptomycin controls surface-borne blight bacteria but will not control internally-borne blight bacteria.

Chickpeas

Several seed treatment fungicides are registered for use on chickpeas. Fludioxonil (Maxim) gives broad spectrum protection against soil-borne fungi; mefenoxam and metalaxyl (Allegiance) provide very specific and highly effective protection against *Pythium* seed rot and seedling blight and early season *Phytophthora* root rot.

Chickpea growers in some parts of western North Dakota have experienced severe difficulty with seed rot of untreated seed and should treat seed with mefanoxam or metalaxyl.

The most serious foliar disease of garbanzo beans is *Ascochyta* blight. *Ascochyta* can be seed-borne, usually at low levels. Wet weather may cause rapid spread of *Ascochyta* from a few infection centers. Thiabendazole (TBZ) is registered for use on garbanzo beans for eliminating seed-borne *Ascochyta.* Purchase of western grown seed that has been laboratory checked for *Ascochyta* also is desirable to minimize the danger of losses from this serious disease, but it should also be treated with thiabendazole.

Lentils

Lentils should be treated with captan, fludioxonil (Maxim), mefenoxam (Apron XL) or metalaxyl (Allegiance). Captan and fludioxonil provide broad spectrum protection against *Rhizoctonia* and *Fusarium* seedling blights. Mefenoxam and metalaxyl (Allegiance) provide excellent control against *Pythium* seedling blight, but do not provide protection against *Rhizoctonia* or *Fusarium.* The two most serious diseases of lentils are anthrac-nose and *Ascochyta* blight. Both can be seed-borne at low levels but seed transmission to developing seedlings has not been deomonstrated for anthracnose. Ascochyta can spread rapidly in wet weather from a few infection centers.

The *Ascochyta* that infects lentils is a different species than the one that infects chickpeas. No fungicide currently registered is effective for elimination of either of these disease fungi on the seed. Purchase of western grown seed that has been laboratory tested for *Ascochyta* is desirable to help minimize the danger of losses.

Flax

Flax seed can be attacked by seed- and soil-borne pathogens, especially when the weather is unfavorable for germination and growth or when the seed coats are cracked and split. Seed treatment with protectant fungicides captan, mancozeb, maneb or thiram will reduce the amount of seed rot and seedling blight. Yellow-seeded varieties are more susceptible to seed coat damage than are brown-seeded varieties. All varieties currently being grown are brown-seeded varieties except Omega, which is yellow-seeded.

Peas

Common seedling blights of peas can be controlled with captan, fludioxonil, PCNB or thiram. The water mold fungi *Pythium* and *Phytopthora* are common problems on peas and can be controlled with mefenoxam, metalaxyl or oxydixyl treatments. No seed treatment is registered for Aschochyta blight on peas, but data from Manitoba indicates that thiram provides good suppression of seed-borne Ascochyta.

Safflower

Safflower rust is borne on the surface of the seed and produces infections on the hypocotyl of the emerging seedlings. Seed-borne rust can result in poor stands and reduced vigour of the seedlings. Fungicidal seed treatment of safflower with one of several available fungicides is recommended to control seed-borne safflower rust. Carboxin, mancozeb, and thiram are labeled for seed treatment use on safflower.

The winter spores of safflower rust can survive from one season to the next in the soil but will not survive to the second season. Control of safflower rust requires seed treatment with a fungicide and crop rotation. Never plant safflower on land that had safflower the year before.

Sugarbeets

Pythium, Aphanomyces, and *Rhizoctonia* are fungi that may cause stand establishment and seedling disease problems in sugarbeets. The *Pythium* fungus occurs in most sugarbeet soils. It causes seed rot, pre-emergence damping off, and post-emergence damping off.

Post-emergence damping off caused by *Pythium* may occur when the seedlings are so tiny that they dry up and blow away within a day or two. Consequently, Pythium-induced seedling death is seldom noticed by the grower. The only thing noticeable may be an unusually poor emergence and stand.

Aphanomyces and *Rhizoctonia* cause death of seedlings at a later growth stage, with plants progressively dying from the two to the eight leaf stage. Later in the season, both fungi may cause a root rot which weakens the plant and also reduces the weight and quality of beet roots. Both diseases are favored by warm soil conditions. *Aphanomyces* is favored by heavy soils with poor drainage, resulting in saturated or puddled soils. *Rhizoctonia* is favored by moist soils.

Most sugarbeet seed is sold treated, but different treatments vary in their effectiveness against these three fungi. Certain fungicides have specific activity: mefenoxam, metalaxyl (Allegiance) and oxadixyl (Anchor) are highly effective against *Pythium;* thiram is moderately effective against *Pythium;* PCNB, chloroneb and fludioxanil (Maxim) are effective against *Rhizoctonia.*

Growers planting in fields with known *Rhizoctonia* problems may wish to request a special or supplemental seed treatment from their seed supplier or else use a planter box overtreatment. Hymexazol (Tachigaren) is effective against *Aphanomyces* and *Pythium.*

Canola

Blackleg is a fungus disease of canola that can cause severe losses. It is seed-borne and also spread by wind-borne as well as rain-splashed spores. The wind-borne spores come from canola crop refuse. Most long distance spread of the blackleg fungus is on seed. Most spread within a field or between fields is from wind-borne spores. The severe (highly virulent) strain of blackleg is common. Benomyl seed treatment provides excellent control of the seed-borne phase of blackleg. Registration of other seed treatments effective against blackleg may occur in 2000 or 2001. Seed treatment is highly recommended for all canola seed planted in areas that do not yet have the severe strain of blackleg. In areas where the severe strain is prevalent, it may be necessary to plant tolerant varieties of canola. Other seedling blights may be controlled with captan, fludioxanil or thiram.

Sunflower

Downy mildew is a soil-borne disease that can cause severe losses if excessive rains occur shortly after planting. The downy mildew fungus is wide-spread across North Dakota. It survives many years in the soil and infects emerging or recently emerged seedlings when the soil is saturated. Several new races have occurred in recent years and at present only one hybrid is resistant to all races.

The downy mildew fungus has developed resistance to mefenoxam, metalaxyl and oxadixyl. Seed treatment with these products provides poor to fair control of downy mildew, depending on what per cent of the downy mildew population is resistant. No suitable replacement fungicide was available for the 2000 growing season.

BIOLOGICAL CONTROL

Kodiak

Kodiak concentrate contains *Bacillus subtilis* bacteria which colonize the developing root system, suppressing disease organisms such as *Fusarium, Rhizoctonia, Alternaria* and *Aspergillus* that attack root systems. When used with a chemical seed treatment, the combination of chemicals and Kodiak provides protection to the root for a much longer time than with chemicals alone.

As the root system develops, the bacteria grow with the roots extending the protection throughout the growing season. As a result of this biological protection, a vigorous root system is established by the plant, which often results in more uniform stands and greater yields. Registered for seed and pod vegtables, soybeans, wheat and barley, and corn plus all other agricultural seeds.

9

Seed Propagation of Woody Ornamentals

For many years seed propagation was the principal method of producing new plants of many woody ornamentals. It was the least expensive means of propagation to produce a large number of new plants from a minimum of stock material. Seed propagation is still used frequently to grow many plants that cannot be propagated asexually. The main disadvantages of seed or sexual propagation are that it fails to produce plants true to variety and that it takes a longer time to produce a salable plant. When propagating woody ornamentals from seed, it is important to collect fruit as soon as it is ripe and before seeds have been dispersed. Seed of some plants will germinate readily if it is harvested immediately after fruit ripens, but if it is harvested after seeds have dried on the tree, pretreatments may be necessary.

SIGNS OF MATURITY VARY

There are no set rules to determine when seeds of selected species are mature and ready for collection. Changes in the physical appearance of fruits—size, shape, weight and colour—can serve as visual guides to seed maturation. As an example, fruits of southern magnolia are a brilliant red when mature and fruits of most junipers change to a deep blue at maturity. Viable seeds of many ornamentals fail to germinate immediately when placed under conditions considered optimum for germination. Such seeds are said to be dormant. However, dormancy is relative because conditions restricting germination vary widely by species. Dormancy is normally the result of the interaction of environmental conditions and hereditary properties of plants. Under various conditions, either the hereditary properties or environmental conditions can predominate and prevent germination. Many woody ornamental plants grown in Florida produce seed that exhibit no dormancy and will germinate readily as soon as fruit are mature and harvested. However, viability of many seed is often very short, sometimes only 3 to 10 days. This situation is especially true for seed coming from pulpy or fleshy fruit. Such seed should be planted immediately after harvest because they lose their viability if stored.

HANDLING SEED FROM FLESHY FRUITS

Fleshy fruits include most palm species, southern magnolia, ardisia, podocarpus, ochrosia sea grape and carissa. When storage cannot be avoided, seeds should be separated from pulp as soon after collection as possible to avoid damaging fermentation. Seeds of species with thin flesh, such as magnolia, can be airdried and planted with seed coats intact. After an initial cleaning or washing such fruits should be spread out in thin layers and dried in the sun or a warm room. An occasional stirring is helpful. Flesh from fruits of palms, sea grape and carissa can be removed by hand or by any of several macerating machines. Residue and seeds may then be separated effectively by flotation in water. Direct a stream of water from a hose at an angle to create a rotary swirl and lifting effect. Empty or nonviable seed, pulp and other debris will either float or sink more slowly than sound, viable seeds. Debris will float to the surface and spill over the edge of the container as the water overflows. Slight stirring of material in the bottom of the container is required. Similar floation methods are used to separate poor from good seeds that have no flesh,

After separation, wet seeds should be surface dried or fully dried in the sun or indoors if they are to be stored. A sieve can be used for finalcleaning to screen or blow away remaining debris. Seeds from dry fruits such as redbud, pines and junipers require only cleaning before planting or storage. Cleaning is simply separation of dry seeds from pods, capsules, or cones, and removal of wings and other appendages. Whether seed are planted immediately after harvest or stored, they should not be treated harshly because they are live plant material. Handle bags or containers of fruits and seeds gently and avoid rough cleaning procedures. Leave seed to dry in heated air only the minimum length of time. Use the lowest possible air temperature to accomplish this job because excessive heat can injure seeds and reduce viability.

FACTORS INFLUENCING LONGEVITY IN STORAGE

Many factors influence longevity of seed in storage, including type of seed, stage of maturity, viability, moisture content and air temperature. However, a few generalizations can be made about storage.

Fully-ripened seeds remain viable longer than seeds collected when immature. Seeds with hard impermeable seed coats store better than those with fleshy coverings.

Fluctuations in temperature and moisture are less favorable than constant conditions. Recommended storage temperature and moisture requirements vary for seeds of woody ornamentals. A rule of thumb applied to agriculture seeds states that conditions for long-term storage are good if the sum of degrees F and percent of relative humidity equals 100 or less. This should serve as a rule for storing tree and shrub seeds. Seeds of many different plants are stored in airtight containers at 40°F or 5°C and 32 percent relative humidity. Oak (*Quercus*) seeds have remained viable for 2 1/2 years stored in polyethlylene bags at 5°C. Seeds can be stored satisfactorily in metal cans, plastic bags, and paper- or aluminum-foil-lined envelopes. A few plants grown from seed by nurseries in Florida are dormant because of a hard seed coat. Seed coats may be impermeable to water, air or both. Seeds, which have hard and impermeable seed coats, include Jerusalem-thorn (*Parkinsonian aculeata*), redbud (*Cercis canadensis*), sweet acacia (*Acacia farnesiana*), *Poinciana spp*. and other legumes. The usual method of overcoming this is by mechanical or acid scarification and soaking in hot water. Scarification is any process to break, scratch or alter the seed coat and make it permeable to water and air.

Mechanical Scarification

Mechanical scarification is accomplished by tumbling seeds in containers lined with sandpaper or other abrasive materials or by mixing seeds with coarse sand or gravel in a revolving container. Small lots of seed can be scarified by rolling them on a cement floor using a brick or board. To determine whether seeds are properly scarified, a test lot can be germinated. The seeds may be soaked to observe swelling, or the seed coats may be examined with a hand lens.

Seed coats generally should be dull but not deeply pitted or cracked enough to expose the inner parts of the seed. Scarified seed will not store as well as comparable nonscarified seed and should be germinated as soon as possible.

The chief advantages of mechanical scarification are as follows:

It requires no temperature controls.

It involves no safety hazards to workers.

Seeds remain dry and can be planted immediately.

The disadvantages are as follows:

Special equipment may be necessary.

Seeds must be free of pulp flesh.

Damage from overtreatment is likely.

Seeds mechanically scarified do not store as well as those scarified with acid or water.

SOAKING SEEDS IN WATER

Soaking seeds in water may overcome seed coat dormancy and stimulate germination in some cases. Impermeable seed coats can be softened by dropping seeds into 4 to 5 times their volume of hot water (80-100°C or 190-210°F).

Heat should be removed immediately and seeds allowed to soak in the gradually cooling water for 12 to 24 hours. Unswollen seeds can be retreated or subjected to some other method of treatment. Seeds should be planted immediately after the hot water treatment. Boiling seeds or over-exposing them to high temperatures is likely to result in injury.

ACID SCARIFICATION

One of the most common methods of pretreating seeds with hard and impermeable seed coats is to soak them in concentrated sulfuric acid. This treatment is highly effective with many species. Special materials and equipment required include:

- Concentrated sulfuric acid to cover the seeds (use commercial grade, which is 95% pure);
- Acid-resistant containers of thick plastic or glass;
- Wire containers and screens for handling, draining and washing the seeds;
- A supply of running water;
- A safe place to drain away the dilute acid after rinsing the seeds; and
- Facilities for drying the seeds if they are to be stored.

Steps in Acid Treatment

Steps in acid treatment are as follows:

Allow seeds to come to air or room temperature, especially if they have been in cold storage.

Thoroughly mix seeds to be treated as one lot.

Determine the optimum period for immersion in the acid. Time of treatment may vary from as little as 10 minutes for some species to as much as 6 hours for other species. Time can be determined on small lots of seeds in preliminary tests by removing samples at set intervals and visually checking thickness of seed coats. When coats become paper thin, treatment should be terminated immediately. Immerse dry seeds in acid for the required period, making sure that all are covered. Usually one part seed to two parts acid will be adequate. Seed should be gently stirred during the immersion period. Treatment should be carried out at 18-27°C or 65-80°F. If temperatures are lower the seeds must be soaked longer; if higher they must be soaked for a shorter time. Remove seeds from the acid and wash promptly and thoroughly over a wire screen in cool, running water for 5 to 10 minutes to remove acid residue. Apply plenty of water at the start and stir carefully during rinsing.

Seed can be planted immediately while wet.

Precautions to Follow in Using Acid

Precautions to take in use of acid are as follows:

Concentrated sulfuric acid is very caustic to skin and to clothing and should be handled with great care.

Never pour water into concentrated sulfuric acid. It will react violently by heating, boiling and splattering.

Several advantages of the acid treatment over mechanical scarification are:

- It is effective for many species.

- It requires little special equipment.
- Cost is reasonable. Acid can be reused.
- There are also disadvantages. They are:
- Length of treatment must be carefully determined.
- Temperature must be controlled.
- Workers face a safety hazard.

Cold Stratification

Seeds of many other species of woody ornamentals such as holly (*Ilex spp.*), southern magnolia (*Magnolia grandiflora*), nandina (*Nandina domestica*), and sweet gum (*Liquidambar styraciflua*) have embryos or endosperms that are nonfunctional, or that contain inhibitors at the time of seed maturity, and require a period of cold stratification. Cold stratification is the subjection of moist seeds to low temperatures for a specified length of time before germination.

Steps in cold stratification are:

Soak seeds overnight or from 12 to 24 hours in water at room temperature immediately before stratification. Mix moist seeds with the moistened but sterile medium. Suitable media include well-washed sand, peat moss, shredded sphagnum and vermiculite. The medium must retain moisture. Seed should be mixed with 1 to 3 times their volume of the medium or placed in layers 1/2 to 3 inches thick alternating with equally thick layers of the stratification medium. Layers of cheesecloth may be used to separate the seeds and the medium. The cheesecloth eliminates the need for cleaning resulting from mixing of the seed and medium. Suitable containers for stratification are flats, trays, boxes or cans that provide aeration, prevent drying and allow drainage. Polyethylene bags no more than 0.004 inch thick have also been used successfully. During stratification seeds should be examined periodically; if dry, the medium should be remoistened.

Place seeds in refrigerated storage. The recommended temperature is between 3-5°C or 37-40°F. For most species of seeds, the required period of low temperature is 3-4 months, but some species such as southern magnolia may require 5-6 months. Plant seeds immediately after removal from refrigeration or they may return to a dormant state, require additional temperature, and lose viability.

COMBINATIONS OF PRE-GERMINATION TREATMENTS

The purpose of combining two or more treatments is to overcome double dormancy, which may result from a combination of hard seed coats, immature embryos or other factors. The combination of scarification with cold stratification is effective for many of these seeds. Also, a combination of warm and cold stratification is another suitable treatment. Prepare seeds for warm stratification using essentially the same procedure as for cold stratification.

Planting seeds directly in a greenhouse for the desired time will suffice for the warm period. Periods of cold temperature must follow the warm temperature.

ENVIRONMENT FOR GERMINATION

Mediums used for germinating seed must have a high water-holding capacity, good drainage and good aeration. A mixture of peat moss and builder's sand in a 1:1 volume is used by many Florida nurseries. Various other materials such as shredded sphagnum moss, vermiculite and perlite have also been used successfully in germination mixtures. Regardless of the media that is used for germination, it should be sterile to prevent loss of seedlings by disease. A common disease problem with seedlings is damping-off, which eventually can cause death of the new plant. Damping-off is caused by certain fungi, primarily species of *Pythium* and *Rhizoctonia*. Control involves three separate procedures: (a) elimination of the organisms during propagation; (b) control of environmental conditions during propagation; and (c) keeping the propagation area sanitized at all times. Excellent control can be obtained by using sterile media, treating seeds and following good sanitation practices. Five percent solutions of commercial bleaching preparations have been used satisfactorily to treat seeds.

Structures for Planting

Seeds may be planted in flats, pots, greenhouse benches and similar structures. Cover germination facilities with moisture-proof plastic, glass or other material to prevent water loss from the medium and surrounding air. Plastic coverings are most desirable because they maintain high humidity without restricting movement of air or oxygen. Optimum temperatures for germination of seeds of woody ornamentals grown in Florida is 24-35°C, or 75-95°F. Variation of 5°C between night and day temperatures often stimulates germination of many species, compared to a constant temperature. Lower the temperature during the dark period. Another critical problem in seed germination is depth of planting. Generally, seeds should not be planted deeper than 1 to 2 times their diameter. Very fine or small seeds should be scattered over the surface of the medium or planted thickly in rows. For small seeds this means dusting them onto the medium surface without covering them, or covering them with a very thin layer of shredded sphagnum moss. Plant larger seed at depths measuring less than their diameter because a 2- to 3-inch planting depth is the maximum for any species. Coconut seeds are an exception.

Importance of Moisture

Moisten the medium before sowing seeds. After sowing the seeds, water with a fine mist or, particularly with small seeds, subirrigate. Both the medium and the air surrounding the germinating seed must remain moist throughout the germination process. Germinating seedlings should receive adequate light to produce short, stocky plants rather than weak and spindly ones. However, full sunlight during very early germination should be avoided with some seedlings because of possible injury from high temperatures. Seedlings should be transplanted into larger containers as soon as the first 2 to 4 tine leaves appear (not the cotyledonary leaves). During transplanting, it is desirable to retain as much of the medium around the roots as possible.

PROPAGATION OF PALMS BY SEED

Follow these eight steps in propagation of palms by seed:

Plant palm seeds as soon as they are ripe. Seeds of some species are relatively short-lived and begin to lose viability in 2 to 3 weeks or less.

Remove seeds' fleshy coats, dry them in the shade for a few days before storage or plant them immediately. Seeds that have been previously dried should be soaked in water for 2 to 3 days before planting.

Before storage, dust seeds with a fungicide, such as ferbam, ziram or chloranil. To hasten germination, scarify or cut through the thick or hard coat of the seeds. Plant the seed in sterilized media in flats or beds. Mixtures of peat and sand 1:1 by volume or peat, sand, and vermiculite 1:1:1 by volume have proven satisfactory.

Keep the seed bed moist and at a temperature of more than 5°C until germination. Germination is considerably better at high temperatures, 30-35°C or 85-95°F. Heating cables may be used to raise the temperature of the seedbed. Germination time of palm seeds varies with species. *Chrysalidocarpus lutescens* and *Phoenix roebelenii* germinate in about one month, whereas *Chamaedora spp.* require about seven months. Transplant seedlings as soon as leaves are well developed to reduce root injury in transplanting.

10

Seed Storage Conditions

In common with all other living things, seeds are subject to ageing and, eventually, to death. In the case of orthodox seeds, the process of ageing and deterioration is so greatly affected by the conditions of storage that the 'age' of seeds, expressed solely in terms of the period elapsed since ripening and harvesting, is an inadequate measure of the degree to which they have 'aged' in the sense of losing viability and progressing towards the irreversible deterioration of death.

The term 'physiological age' is commonly used to describe the degree of deterioration of seeds measured by their reduced capacity for germination. Nomographs for the effects of temperature and MC on physiological ageing of seeds have been constructed for several agricultural crops. As an example, the nomograph for barley indicates that the same degree of deterioration (from initial 95 per cent to final 50 per cent germination) would occur in about 16 days in seeds stored at 25°C and 21 per cent MC, but in about 100 years in seeds stored at 8°C and 8 per cent MC. Both seed lots would have the identical physiological age, though stored for very different periods of time. Similar effects can be expected in orthodox seeds of forest trees.

A number of physiological changes in cell tissues may be associated with physiological ageing in seeds.

They include:

- Loss of food reserves caused by respiration, *e.g.* decrease in proteins and non-reducing sugars, accompanied by increase in reducing sugars and free fatty acids
- Accumulation of toxic or growth-inhibiting by-products of respiration
- Loss of activity of enzyme systems
- Loss of ability of dried proteid molecules to recombine to form active protoplasmic molecules on subsequent rehydration
- Deterioration of semi-permeable cell membranes
- Lipid peroxidation, leading to production of free radicals which react with, and damage, other components in the cell

- Alterations to DNA in cell nucleus, causing genetic mutations as well as physiological damage. It is still uncertain to what extent these various effects are the causes or only the symptoms of deterioration, but it has been suggested that the production of free radicals is the first effect of ageing and that damage to the several systems in the cells is the subsequent result of the release of free radicals.

Whatever the exact mechanism of seed deterioration, there is a consensus that, in orthodox seeds, loss of seed viability is largely governed by the rate of respiration. Any measures which reduce the rate of respiration without otherwise damaging the seed are likely to be effective in extending longevity during storage. These are the control of oxygen, the control of moisture content and the control of temperature. Inrecalcitrant seeds the safe minimum levels of oxygen, moisture content and temperature, and hence of respiration, are all considerably higher than those for orthodox seeds but, provided levels are maintained above the safe minima for each species, it appears that longevity can be extended by keeping them as close to the minima as possible in order to avoid an excessively high respiration rate.

SEED STORAGE CONDITION

Even in ideal storage conditions seed will soon lose viability if it is defective from the start.

Factors to be considered are:

- *Seed maturity*: Fully ripened seeds retain viability longer than seeds collected when immature. Certain biochemical compounds, essential for preserving viability, may not be formed until the final stages of seed ripening. These include dormancy-inducing compounds in certain species, and dormancy is sometimes associated with seed longevity. In a few species *e.g.* Gingko biloba,Fraxinus excelsior, seed embryos are underdeveloped when the seed is shed. Maturation of these embryos is necessary before sowing but need not be done before storage. In Fraxinus excelsior drying of freshly collected samaras to 9–10 per cent MC, followed by storage in sealed containers at-3° C, gives satisfactory results provided that successive moist warm and moist cold treatments are applied after storage.
- *Parental and annual effects*: In seed harvest, quantity and quality often go together. The percentage of sound seeds in a high-yielding mother tree is usually higher than in one with a poor crop. Similarly, a given mother tree will have a higher percentage of sound seeds in a good crop year than in a poor one. Collection from high-cropping mother trees in a seed year is likely to yield seeds with the best longevity in storage. On the other hand, high-cropping 'wolf' trees should be avoided because of their potentially

undesirable wood properties, even though they may produce long-storing seeds.

- *Freedom from mechanical damage*: Seeds damaged mechanically in extraction, cleaning, dewinging etc. rapidly lose viability. The danger is greatest for species which have thin or soft seedcoats. Excessive heat during extraction or drying also damages seed. Care should be taken to use the minimum times, lowest temperatures and minimum machine speeds necessary during the preparation of seed for storage. In some species, damage during dewinging may be reduced by partly restoring the moisture content between extraction from cones and dewinging, since moist seeds suffer less mechanical damage than dry ones.
- *Freedom from physiological deterioration*: Poor handling in the forest, during transit or during processing causes physiological deterioration of seeds even if mechanical and fungal damage are absent. Adequate ventilation of orthodox seeds is necessary to avoid rapid respiration and overheating, while recalcitrant seeds must be protected against excessive drying.
- *Freedom from fungi and insects*: For species stored at low temperature and low moisture content, the storage conditions themselves should prevent the development of fungi and insects. It is necessary, however, to avoid collection of crops showing a high incidence of fungal or insect attack and to carry out all operations of collection, transport, processing etc. as quickly as possible to ensure seed is not already damaged before it goes into storage. Attack by fungi and insects is most rapid on the forest floor, so collection from the ground should be carried out as soon after fruit fall as possible. Fungicidal treatment cannot be generally recommended since it can be harmful to seeds; many fungicides are only effective when dissolved in water and are inappropriate for dry storage. Insects are usually killed if seeds are dried at temperatures above 40°–42° C. For seeds which cannot be dried, other measures may be needed. For example seeds of Quercus are fumigated with serafume or other chemicals or heated in warm water for control of weevils, while methyl bromide or carbon bisulphide are also commonly used to kill insects.
- *Initial viability*: Seed lots with high initial viability and germinative capacity have a higher longevity in storage than those with low initial viability. Germination tests, preceded if necessary by appropriate pretreatment to overcome dormancy, should be carried out on a sample of each seed lot before storage, in order to determine how long the seed is likely to retain viability in storage. Longevity of the viable seeds is correlated with the percentage

which germinate in the initial test. As an example, samples of two seed lots of the same species, from which 80 per cent germination of fresh seeds is normally expected, might give results of 90 per cent and 50 per cent initial germination. Not only would storage of the second seed lot involve wasting space in storing dead seed, but even the 50 per cent of initially viable seeds are likely to lose their viability more quickly in storage than would the 90 per cent viable seeds in the first seed lot. Deterioration in initial viability may not be serious if the seeds are to be sown within a few weeks or months, but only good quality seed should be stored for long periods. In long-term storage of agricultural seeds for genetic conservation, it is recommended that no seeds should be accepted for storage which have an initial viability of less than 85 per cent of that considered typical for the species or variety in question. It may be noted that initial viability and germinative capacity are frequently the resultant of the factors described in previous paragraphs (seed maturity, mechanical damage, fungal or insect attack).

SEED STORAGE METHOD

A number of different storage methods are available, as described below. The main factors affecting choice are the seed characteristics of the species in question, the period for which it is to be stored and the cost. If more than one method is suitable to maintain viability for the required period, the simplest and cheapest will normally be chosen.

STORAGE AT AMBIENT TEMPERATURE AND HUMIDITY

Seeds can be stored in piles, single layers, sacks or open containers, under shelter against rain, well ventilated and protected against rodents. Best results are obtained in cool, dry climates. In these conditions several species of Pinus, Eucalyptus Pseudotsuga and Tectona will store satisfactorily for at least six months, while leguminous trees with impermeable seedcoats and naturally low MC, *e.g.* Acacia Prosopis, Robinia will retain viability for years.

DRY STORAGE

Orthodox seeds will retain viability longer, when dried to a low moisture content (4–8 per cent), as described, and then stored in a sealed container or in a room in which humidity is controlled, than when stored in equilibrium with ambient air humidity. Storage life is further prolonged if cool, but not controlled, temperature conditions can be provided, *e.g.* at high latitude or altitude and in a cellar or other room screened from direct sunlight.

Seed is sometimes stored in open containers in a room maintained at an RH of 15–20 per cent by dehumidifying machinery. In forestry it is more common to rely on predrying of the seed to the correct MC, followed by storage in full, sealed containers. Provided that the containers are not opened too frequently and that the sealing is effective, the method will maintain a low MC for many years. It is cheaper than using a humidity-controlled room, especially during periods when only a little seed is in store, and it is not subject to hazards of mechanical breakdown.

This method is suitable for a range of species including many Pinus and Eucalyptus species, for which it should maintain viability for one or more years.

CONTROL OF BOTH MC AND TEMPERATURE

This storage treatment is the standard practice for many orthodox species which have periodicity of seeding but which are planted annually in large-scale afforestation projects. For many species a combination of 4–8 per cent MC and 0 to +5°C temperature will maintain viability for 5 years or more. Some cool-temperate genera benefit by storage at sub-freezing temperatures, *e.g.*-4°C or lower for Abies,-10°C forFagus,-18°C for Pinus strobus, Populus deltoides and others. Pinus merkusii is an example of a tropical pine which responds well to storage at low temperature and MC. 80 per cent germination was obtained after 3 years' storage at 2°C and 6–10 per cent MC of seeds of the Zambales (Philippines) provenance, while seeds stored in equilibrium with room temperature and humidity showed a significant loss of germination after 3–4 months. P. caribaeaand P. oocarpa behave similarly in this respect.

In addition to true seeds, this method is also suitable for certain types of fruits. For example in the Jari project of Brazil depulped, cleaned and dried stones of Gmelina arborea are successfully stored in sealed containers at 5°C and 6–10 per cent MC. The fresh stones have a germination of 90 per cent and after two years of storage germination is still 80 per cent.

DRY STORAGE FOR LONG-TERM GENE CONSERVATION

The preferred storage treatment for long-term conservation of gene resources of orthodox agricultural seeds is-18° C temperature and 5 per cent +-1 per cent MC. This is likely to be equally appropriate for orthodox seeds of forest trees requiring storage for genetic conservation. The quantity of seeds requiring this standard of storage is small in comparison with the quantities used each year for operational afforestation, and the cost per kg of seed is higher.

For many countries it would therefore be desirable that forest and agricultural crop genetic resources should share a common long-term storage facility. A good example is that of the Banco Latino Americano de Semillas Forestales at CATIE, Turrialba which has its own seed store (55 m^3 capacity at 5°C) for short or medium term storage, but also has access to the long-term storage facilities (at-20°C) of the Regional Genetic Resources Unit, which are also at Turrialba. Loss of viability in storage, in addition to reducing the number of plants which can be produced by a given seed lot, may result in a shift in the genetic constitution of the seed being stored. This could be particularly important in forest trees which are predominantly outbreeding, variable populations. First, loss of viability may occur more rapidly in some genotypes than others; if losses are high, say 50 per cent of the total, the genotypes with short-lived seeds may be eliminated altogether. Yet they may have valuable traits for adaptation, growth or disease resistance as growing trees, and in any case they contribute to genetic variation in the species which it is the purpose of genetic conservation to preserve. Secondly, it is an accepted fact in agriculture that chromosome damage or change occurs and accumulates in the seed in storage, and that the risk of such heritable gene mutations depends not so much on the age of the seed as on changes in its viability.

A seed lot which has suffered a serious loss of viability is likely to have experienced some gene mutations among the survivors, but there is very little direct evidence that heritable mutations are induced under good storage conditions which lead to only small losses in viability.

The high standard of storage conditions recommended by the IBPGR and referred to above, if combined with regular testing of seed and regeneration as soon as germination falls to 85 per cent of the initial germination rate should minimize the risk of genetic change in storage. It is possible that still lower temperatures would increase longevity even more. Research on storage in liquid nitrogen has been pursued for some years and considerable progress has been made, but testing for several more years will be needed before the method can be recommended for general adoption in gene banks.

MOIST STORAGE WITHOUT CONTROL OF MC OR TEMPERATURE

Suitable for storage of recalcitrant seeds for a few months over winter. Seeds may be stored in heaps on the ground, in shallow pits in well-drained soils or in layers in well ventilated sheds, often covered or mixed with leaves, moist sand, peat or other porous materials. Seeds stored outdoors are kept moist by rain or snow, but those under shelter may need to be moistened periodically. The aim is to maintain moist and cool conditions, combined with good aeration to avoid overheating which may result from the relatively high rates of respiration associated with moist storage. This may be accomplished by regular turning of the heaps of seed or by inserting bundles of straw or twigs into them.

This method is suitable for short-term storage of large-seeded hardwood species in the temperate zone *e.g.* Quercus, Castanea, Aesculus. It is unlikely to be suitable for tropical recalcitrant species because ambient temperature is too high.

Outdoor stratification, a method of overcoming internal dormancy, is described. It is properly to be considered as a seed pretreatment, but it serves the incidental function of storing the seed for a few weeks or months and the method used is closely akin to those described.

MOIST COLD STORAGE WITH CONTROL OF TEMPERATURE

This method implies controlled low temperatures just above freezing or, less commonly, just below freezing. Moisture can be controlled within approximate limits by adding moist media *e.g.* sand, peat or a mixture of both to the seed, in the proportions of one part media to 1 part seed by volume, and remoistening periodically, or more accurately (but more rarely) by controlling the relative humidity in the cold store.

The latter type of control is often too expensive. Respiration rate is reduced and storage life prolonged by the low temperature, but seed should not be stored in sealed gas-proof containers which would limit oxygen supply. Closed polyethylene bags of 4–10 mil (100–250 microns) thickness will allow exchange of oxygen and CO_2 with air outside, while severely restricting exchange of moisture.

The method is suitable for the same temperate recalcitrant genera as listed in the previous section, and with temperatures of 0–5° C should extend viability up to 1½ to 2 years. Sub-freezing temperatures have given improved results in a few cases but frequently injure seeds with high MC and should be used only after research has demonstrated their applicability to the species in question.

Less is known about the application of this method to tropical species, but it merits much more investigation that it has so far received, for the dipterocarps and genera such as Araucaria, Agathis and Triplochiton. As mentioned earlier there is evidence that some species are killed at low but above-freezing temperatures and Gordon has proposed a division of recalcitrant seeds into those which can withstand temperatures below 10° C without loss of viability and those which cannot.

Tamari, summing up several years of research on dipterocarps in Malaysia, concluded that the best treatment for several species was:

- Dry at a temperature not above 35°C, to reduce MC to 35 per cent
- Seal with a fungicide inside polythene bags
- Store at 15°C, or for 3 weeks at 15°C followed by further storage at 10°C.

This treatment has been successful in extending longevity from a week or two up to two months in Hopea helferi, but this is still a long way from providing safe storage between seed years, reported to vary from 3–6 years apart in many dipterocarps. Storage at 3.5°C and MC over 32 per cent over periods of at least 6 months has been successful on the recalcitrant Araucaria hunsteinii.

In some recalcitrant species newly germinated seeds may retain viability better under moist cool storage conditions than ungerminated seeds. Gordon reported that some pregerminated seedlots of Quercus spp. showed no significant change in the number of living seedlings after one year's storage in 500 gauge (125 microns) polythene bags lightly sealed at 3°C, whereas a large proportion of the seeds which were viable but ungerminated when placed in the same bags died during the same period.

OTHER METHODS

Other storage methods have been used in the past, but are not yet of wide application.

They include:

- Storage of recalcitrant seeds in running (not stagnant) water.
- Storage under partial vacuum.
- Storage in gases other than air *e.g.* nitrogen or CO_2.
- Coating individual large seeds with paraffin or latex to prevent moisture exchange. This method may also be used to maintain moisture content during shipment.

PURPOSE OF SEED STORAGE

Seed storage is the maintenance of high seed germination and vigour form harvest until planting. Is important to get adequate plant stands in addition to healthy and vigourous plants. Every seed operation has or should have a purpose. The purpose of seed storage is to maintain the seed in good physical and physiological condition from the time they are harvested until the time they are planted. Seeds have to be stored, of course, because there is usually a period of time between harvest and planting. During this period, the seed have to be kept somewhere. While the time interval between harvest and planting is the basic reason for storing seed, there are other considerations, especially in the case of extended storage of seed.

Seed suppliers are not always able to market all the seed they produce during the following planting season. In many cases, the unsold seed are "carried over" in storage for marketing during the second planting season after harvest. Problems arise in connection with carryover storage of seed because some kinds, varieties, and lots of seed do not carryover very well.

Seeds are also deliberately stored for extended periods so as to eliminate the need to produce the seed every season. Foundation seed units and others have found this to be an economical, efficient procedure for seed of varieties for which there is limited demand. Some kinds of seed are stored for extended periods to improve the percentage and rapidity of germination by providing enough time for a "natural" release from dormancy.

Regardless of the specific reasons for storage of seed, the purpose remains the same maintenance of a satisfactory capacity for germination and emergence. The facilities and procedures used in storage, therefore, have to be directed towards the accomplishment of this purpose.

In the broadest sense the storage period for seed begins with attainment of physiological maturity and ends with resumption of active growth of the embryonic axis, *i.e.*, germination. Seeds are considered to be physiologically and morphologically mature when they reach maximum dry weight. At this stage dry-down or dehydration of the seed is well underway. Dry-down continues after physiological maturity until moisture content of the seed and fruit decreases to a level which permits effective and efficient harvest and threshing. This stage can be termed as harvest maturity. There usually is an interval of time between physiological maturity and harvestable maturity, and this interval represents the first segment of the storage period. Any delays in harvesting the seed after they reach harvest maturity prolongs the first segment of the storage period–often to the detriment of seed quality.

The second segment of the storage period extends from harvest to the beginning of conditioning. Seed in the combine, grain wagon, and bulk storage or drying bins are in storage and their quality is affected by the same factors that affect the quality of seed during the packaged seed segment of the storage period. The third segment of the storage period begins with the onset of conditioning and ends with packaging. The fourth segment of the storage period is the packaged seed phase which has already been mentioned. The packaged seed segment is followed by storage during distribution and marketing, and finally by storage on the farm before and during planting.

The control that a seedsman has over the various segments of the storage period for seed varies from a high degree of control from harvest to distribution, to much less control during the postmaturation-preharvest, distribution-marketing, and on-farm segments. Despite variable degrees of control over the various segments of the storage period, the seedsman's plans for storage must take into consideration all the segments. The things that can be done must be done if the quality of the seed is to be maintained.

METHODS

The conditions which prolong viability during storage have been well defined for seeds which are tolerant of desiccation. Storage conditions have been recommended by the IBPGR Advisory Committee on Seed Storage. For base collections, seeds of between 3-7% moisture content should be stored in sealed containers. Sub-zero temperatures are acceptable, but –18 °C or less is preferred. For active collections sealed storage of seeds dried to 7% moisture content or less is recommended at temperatures of less than 15 °C. Unsealed storage is not encouraged. In particular, it is not recommended in tropical areas.

CHECK THE NUMBER OF SEEDS IN THE ACCESSION

1. Weigh the seeds of each accession.
2. Convert the weight of seeds to seed number, using the thousand seed weight of each accession for an accurate conversion. An approximate conversion can be done on a species basis by using

the approximate weights given in Appendix 2 of Cromarty, Ellis and Roberts (1982).

3. For accessions containing mixtures of genotypes, the sample size should be at least 4000 seeds. For genetically uniform accessions, the sample size should be at least 3000 seeds.
4. If the sample contains more than the required number of seeds, proceed to storage.
5. If the sample contains less than the required number of seeds, either proceed directly to regeneration or store the seeds temporarily in the genebank and regenerate at the earliest opportunity.

Notes and Examples

An approximate inter-conversion of seed number and weight can be done easily using the thousand seed weight.

Example for Sorghum (1000 seed weight = 17 g) 100 g contains:

$$\frac{1000}{17} \times 100 = 5750 \text{ seeds}$$

These are minimum sample sizes for the start of storage and if both space and seeds are available, more seeds should be held.

Regenerate as soon as possible. Seeds processed and stored under good conditions will not loose viability before regeneration.

Equipment

Coarse balance

DETERMINE WHERE THE SEEDS SHOULD BE LOCATED

1. Check the inventory data file of the genebank to find the next available space where a container can be located.
2. When seeds from the same regeneration cycle of the same accession are stored in several containers, keep all the containers of the accession together.
3. Make a list of where each accession will be placed.

Notes and Examples

The storage arrangements will vary among genebanks. The most important point is to know exactly where to locate each accession within the store.

Equipment

- Racks
- Trays, boxes or drawers
- Coldroom or freezer

PLACE IN THE SEED STORE

1. Place the container into the seed store in the listed location.

ENTER THE DATA INTO THE DATA BASE

1. Fill in the data on the location and date of storage of each accession and each container into the data file.
2. Record the date of the next monitoring test for germination in the data file. This date will be determined by the curator after considering the viability and moisture content of the seeds, storage conditions and the IBPGR recommendations.

Notes and Examples

A code can be used to locate an accession within the store. Each unit can be identified by a number or letter in the store. The code can indicate the number or the letter of the freezer, store, rack, basket or drawer, etc.

Example

A010201 could be used to indicate the location as:

Coldroom:	A
Rack:	01
Shelf:	02
Box:	01

Colour codes can also be a quick and easy way to locate accessions. A colour can be used for each rack, shelf or species. This both speeds up the work in the coldroom and makes it easy to spot errors. Owing to the very cold temperatures, the faster that one can locate accessions in the cold room the better.

STORAGE ATMOSPHERE

The most obvious method of reducing the rate of aerobic respiration is to exclude oxygen from the atmosphere surrounding the seeds. This can be done by replacing oxygen by other gases such as CO_2 or nitrogen, or by using a partial or complete vacuum. In an example with lettuce cited by Roberts, seeds were stored in sealed containers at 6 per cent MC and 18° C. After 3 years, seed stored in an atmosphere of pure oxygen had 8 per cent viability, those in air 57 per cent, those in nitrogen or argon or CO_2 78 per cent and those in a vacuum 77 per cent.

The value of excluding oxygen during storage of dry orthodox seeds has also been demonstrated in Pinus radiata. Best results were obtained with a storage atmosphere of nitrogen, followed by CO_2, while vacuum and air both gave poorer results. At the highest temperature used, 35°C, at which deterioration in viability was most rapid, the loss in final germination after

50 weeks' storage in sealed containers at 8 per cent MC was 8 per cent in nitrogen, 14 per cent in CO_2, 21 per cent in vacuum and 29 per cent in air. The same ranking was obtained by comparing the speed of germination and the vigour of germinated seedlings (measured as dry weight 49 days after sowing).

Although increases in seed longevity of this magnitude have been achieved experimentally, some of the methods are expensive to apply and the effects on seed life are less dramatic than the effects of differences in temperature and humidity. Exclusion of oxygen will prevent aerobic, but not anaerobic, respiration, whereas reduced MC and temperature will decrease the level of both. While systematic predictions have been made of seed longevity under a range of temperature and MC for several agricultural crops, similar quantitative predictions of the effect of oxygen levels on longevity are lacking.

One simple method which is recommended is to fill sealed containers as nearly full as possible. If there is only a small amount of air inside the container as compared with the volume occupied by the seeds, oxygen will be consumed and CO_2 produced. The resulting high CO_2/O_2 ratio is probably favourable for seed longevity in orthodox seeds.

Whereas the complete exclusion of oxygen from the storage atmosphere appears beneficial to most dry orthodox seeds, there is evidence that some oxygen is necessary for recalcitrant seeds. Şeeds of Araucaria hunsteinii, which had an initial germination of 56 per cent, had all died within a month if stored in pure nitrogen, in two months if stored in 1 per cent oxygen, and in three months if stored in 5 per cent oxygen, while germination was still 18 per cent after four months' storage in 10 per cent oxygen. Storage in a polythene bag of 25 micron thickness, which was ventilated (21 per cent oxygen) periodically when opened for extraction of samples, gave results very similar to those in 10 per cent oxygen. King and Roberts record a general consensus that adequate ventilation (*i.e.* adequate oxygen), is necessary for the successful storage of recalcitrant seeds at relatively high MC, as well as for storage of imbibed seeds of orthodox species.

STORAGE TEMPERATURE

Temperature, like moisture content, is negatively correlated with seed longevity; the lower the temperature the lower the rate of respiration and thus the longer the life-span of the seed in storage. Harrington suggested another rule of thumb for agricultural seeds-between 50°C and 0°C, every 5°C lowering of storage temperature doubles the life of the seed. For orthodox seeds, which can be dried to a low moisture content, still greater longevity can be assured by storage at sub-freezing temperatures. For long-term storage for genetic conservation of agricultural seeds, a temperature of-18°C has been recommended as the 'preferred' standard for most species and-10°C as an 'acceptable' standard for those species known to have an intrinsic high viability.

Much lower temperatures have been used with success on an experimental basis, *e.g.* in liquid helium at-269°C, but the high cost of maintaining such low temperatures for a long period would outweigh any (as yet unproven) advantages in increased longevity. Choice of storage temperature varies considerably according to species and the period for which the seed is to be stored.

The lower the temperature that has to be maintained in a cold store, the higher the cost, and provision of subfreezing temperatures may be unnecessary if seed is to be stored for only a year or two for afforestation projects. Holmes and Buszewicz noted that, in a number of experiments on conifers, the superiority of sub-freezing temperatures became evident only after prolonged storage over periods of about 5 years or more. Sub-freezing temperatures (-18°C) appear to prolong viability in tropical orthodox species such as Eucalyptus deglupta and Flindersia brayleyana.

Some seeds keep well at room temperature, *e.g.* many leguminous and rosaceous genera, Eucalyptus, Tilia and many other hard-seeded or stone fruits. Most species, however, only keep well for longer periods at lower temperatures. In 3–5 years' storage of most conifers and Alnusand Betula, the crucial temperature seems to lie at a maximum of +4°C. The temperature should therefore be kept at 1 to 4°C. For longer periods of storage, say 5–15 years, the temperature should be-4 to-10°C. For Abies, however, a temperature of-4°C is used for short storage periods and-10° to-20°C for longer periods.

Temperature and moisture factors are so interrelated that it is very difficult to separate them. Seeds at a relatively high moisture level can be stored for considerably longer periods at near freezing temperatures than at higher temperatures, while higher storage temperatures (30°C) are less harmful when the moisture content of the seed is low. In short, it can be said that the critical moisture content lies at a higher level when storage temperatures are low than when they are intermediate or high, *i.e.* to some extent, a low temperature can compensate for a high moisture content, and *vice versa*. It is, however, necessary to avoid any risk of freezing damage caused by ice formation in seeds of high MC. Roberts has suggested that 20 per cent MC may be the critical upper limit for storage at 0°C, 15 per cent for-20°C and 13 per cent for-196°C. If seeds are dried to 4–8 per cent MC as commonly recommended for orthodox species, there should be no danger of freezing damage, even at temperatures well below zero. As mentioned in Chapter 6, the equilibrium moisture content of many seeds at a given RH varies with temperature.

Barton and Crocker have shown that, at a range of RH from 35 per cent to 76 per cent, the amount of water contained by seeds increased progressively as temperature decreased from 30°C to 10°C. Species included Pinus as well as several agricultural crops. At low (35 per cent) RH moisture absorption by dry seeds was approximately the same at 5°C as at 10°C, but at higher (55 per cent and 76 per cent) RH the seeds absorbed less moisture at 5°C than at 10°C, thus reversing the trend above 10°C. Change of equilibrium moisture content with change of temperature can be of importance in open storage. In sealed containers the effect is minimal because the final EMC is dominated by the initial MC of the seeds and not the moisture of the enclosed air.

As with moisture content, repeated fluctuations in temperature lead to loss of viability. As far as possible, temperature should be maintained at a uniform level.

The effect of temperature on seed longevity of temperate recalcitrant species is similar to that of orthodox species-within certain limits the lower the temperature the longer the period of viability. Some tropical species are killed by temperatures above freezing *e.g.* some dipterocarps at <14°C, cacao at <10°C and mango at < 3–6°C. Seeds of Hopea helferi stored at 15°C with high moisture content in unsealed polythene bags retained 98 per cent germination after 37 days and 80 per cent after 60 days. Germination was much reduced if temperature was dropped to 10°C or raised to 25°–28°C. Shorea ovalis is another species which does not withstand low temperatures; it stores best at 21°C. Shorea talura, in contrast, stores well at 4°C and 40 per cent MC (wet weight basis); after six months germination was reduced from an initial 95 per cent to 69 per cent. Other species of dipterocarp have a much shorter seed life.

For most temperate recalcitrant species the lower the temperature down to 0°C the longer the safe storage period. Below-freezing temperatures, on the other hand, often kill recalcitrant seeds which need to be stored at a high moisture content. Some success has been achieved in storing Quercus seeds in the USA by maintaining them at 35–45 per cent moisture content and-1° to +3°C temperature. Temperature can be critical because less than-1°C will usually kill the seeds, while temperatures above 2° or 3°C cause excessive germination. In Europe more northerly species of Quercus can be stored at slightly lower temperature-1° to-3°C and 38–45 per cent MC.

LIGHT

Light, particularly ultra-violet light, is reported to be harmful to seed, but very few studies have been made. Use of opaque metal containers would be preferable to glass jars or bottles for species which are affected by light. But light appears to be much less important than either moisture content or temperature.

RELATIONSHIPS OF SEED MOISTURE CONTENT

The relationships of seed moisture content on a wet weight or fresh weight basis to seed MC on a dry weight basis, and of the equilibrium moisture content of seeds to the relative humidity of the surrounding atmosphere, are important in seed processing and are explained. They are equally important in seed storage. In the first case manipulation of RH can effectively change

MC of seeds to the optimum for storage, in the second case MC can be maintained at or near that optimum by maintaining a suitable RH in the atmosphere around and between the seeds.

Effect of MC. In orthodox seeds, moisture content is probably the most important single factor in determining seed longevity. Reduction in MC causes a reduction in respiration and thus slows down ageing of the seed and prolongs viability. Harrington, cited by Barner, has related MC to various processes within and around the seed as follows:

Prevention of fungal activity is more easily achieved by controlling MC than by controlling temperature. If MC and RH are high enough, fungal activity is possible between-8°C and +80°C and it is easier to keep MC below 12–14 per cent (or RH to the equilibrium of around 65 per cent) than to maintain sub-zero temperatures.

Within the range of 4 to 14 per cent MC, Harrington has suggested a rule of thumb applicable to many agricultural species-the life of the seed is doubled for every 1 per cent decrease in MC, Schönborn found a relationship of a similar order when measuring the respiration rate, expressed in terms of CO_2 production, of Picea abies. At 20°C and 20 per cent MC the seed gave off 80 ml. CO_2 per hour per kg of seed; at 20°C and 5 per cent MC, the rate of CO_2 production was reduced to 0.11 ml/hr/kg, a reduction of nearly a thousand times for a difference of 15 percentage points in MC.

An MC of 4–8 per cent is considered safe for most orthodox species; 5 per cent ± 1 per cent is recommended for long-term storage for genetic conservation. Oily seeds will usually tolerate drying to a somewhat lower moisture content (calculated on the basis of total fresh weight) than will non-oily seeds. Drying below 4 per cent can lead to damage or more rapid loss of viability in some species, although certain species can be dried to a considerably lower MC. Betula papyrifera was successfully stored at 0.6 per cent MC without injury, while Schönborn succeeded in drying small samples of Picea abies, Pinus sylvestris, Pseudotsuga menziesii and Larix decidua down to 0 per cent MC without any observable drop in germination after 6 months, compared with germination at the normally applied 6–8 per cent MC.

Drying in this case was done, not by exposing the seeds to high temperatures but by leading a current of dry air through the seeds at 20° C. The same treatment killed Pinus strobus and Abies alba, and earlier attempts by Barton to store seeds of several species of Pinus and Picea at 0 per cent MC also failed. Below about 2 per cent MC desiccation injury becomes a strong possibility in many species. Drying to very low MC is also more costly than drying to the usual 4–8 per cent and is likely to be used only in exceptional cases. Methods of drying are described on pp. 95–107.

Some orthodox forest trees store best at appreciably higher MC. As mentioned on p. 134, 8–10 per cent is recommended for Fagus sylvatica. For seeds of Abies spp., an MC of 12–13 per cent is recommended for storage of one to three years, but for longer periods this should be reduced to 7–9 per cent. Species which benefit from storage at higher than average MC also need particular care in the timing and speed of drying.

Fluctuation in the moisture content of seed in storage due to open storage without humidity control or to frequent opening and resealing of sealed containers results in deterioration in the germinability of the seeds. In fact a steady MC slightly above the optimum is usually less harmful than one which fluctuates between the optimum and a higher moisture content.

Some cases have been reported in which the usual trend of decreasing seed longevity associated with increasing MC is reversed at or near the moisture content of fully imbibed seeds. If the species in question needs exposure to light in order to germinate, it is possible to store fully imbibed but ungerminated seeds for some time in the dark. Fraxinus americana was stored at 22° C at varying MCs with the following results:

MC %	**Germination (%) After Storage Periods Indicated**			
	1 Month	**2 Months**	**3 Months**	**4 Months**
6.0	98	92	96	94
9.5	94	88	76	4
18.6	81	22	0	0
Fully Imbibed (In Dark)	96	95	98	96

It has been postulated that imbibed seeds can repair damage to cell membranes, enzymes and DNA in the cell nucleus caused by free radicals in a way which is not possible for seeds at lower MC. Prolonged imbibed storage may, however, be difficult in practice, because of the need to maintain constant high moisture for imbibition and adequate oxygen without allowing the seeds to germinate or encouraging the multiplication of fungi and bacteria.

Moisture content is also important in recalcitrant seeds, but in this case the critical MC is the minimum to which it is allowable to dry the seeds rather than the maximum content for prolonged storage. For many of the large temperate hardwood seeds, moisture contents in the range of 25–79 per cent are appropriate. Storage should be carried out at close to the minimum safe MC, since the higher the MC the higher the respiration rate and the more rapid the loss of viability. Higher respiration rates release higher amounts of energy and there is a risk of overheating and death of the seed, unless great care is taken to provide adequate aeration.

High MC also increases fungal activity and the spread of rot. Wang quotes results from two seed lots of Acer saccharinum; germinability of one lot stored at 58 per cent MC and 1–2° C dropped from 94 per cent to 12 per cent after 6 months' storage, while that of the other, stored at 45 per cent MC at the same temperature, was still 78 per cent after 16 months. Loss of viability in this species may be sudden. Tylkowski found that seeds stored in sealed bottles at 50–52 per cent MC and-1° to-3°C had over 90 per cent germination after 18 months but this dropped to nearly zero after 24 months.

Less research has been done on tropical recalcitrant species, but there is some evidence *e.g.* in Triplochiton, that viability may be significantly prolonged if the minimum MC for the species can be determined and if particular care is taken over the period of drying down to that MC. Trials of Shorea platyclados in Malaysia indicated that a gradual reduction of MC to 20–27 per cent, followed by sealing in charcoal, sawdust or vermiculite at 15°–22° C allowed storage for at least one month, compared with the week or so of natural viability.

On the basis of experiments carried out on Shorea parvifolia and Dipterocarpus humeratus, Maury-Lechon *et al* recommended reduction of MC to between one quarter and one half of the initial MC in freshly collected fruits. Although tropical recalcitrant seeds cannot yet be stored for more than short periods, there is a growing body of useful research on the problem. King and Roberts contains a good summary of achievements and possible approaches.

MAINTAINING SEED VIABILITY

Most, if not all, of the species which have been recorded as maintaining seed viability over a period of decades are hard-seeded. They include a number of tropical leguminous species. Examples of species of which at least some seeds maintained viability after lengthy periods of storage in herbaria, cited by Harrington from the work of Ewart and Becquerel are:

158 YearsCassia Multijuga
149 YearsAlbizzia Julibrissin
115 YearsCassia Bicapsularis
99 Years Leucaena Leucocephala

Ambient conditions in herbaria storage can be considered good (fairly low relative humidity and temperature) but well short of the combination of low initial MC, sealed storage and sub-freezing temperature now considered ideal for longterm storage of orthodox species.

Recent research has provided more precise information on conditions of storage, initial germination and final germination of some species, but over shorter periods. Examples are:

As explained later in this chapter, modern thinking has defined low MC, low temperature and low oxygen pressure as the three most important constituents of the storage conditions which man should provide to maximize seed longevity in orthodox species. In the impermeable seedcoat nature has provided two of these constituents, a low MC and exclusion of oxygen. Full-sized but green leguminous seeds, sown immediately without drying, may germinate at once, indicating that the seedcoat has not yet developed an impermeable layer; no doubt the development of impermeability is synchronized in nature with the reduction of seed moisture by natural drying to the optimum content for longevity. Hard seeds are thus a potent factor in extending seed life in all conditions of storage but confer their most important benefits when storage facilities are limited and during the potentially dangerous period between collection and entry into long-term storage.

Not all leguminous seeds are equally long-lived, for example Koompassia malaccensis seeds have thinner seedcoats and deteriorate more rapidly in storage than species such as Parkia javanica, and they need no pretreatment to overcome seedcoat dormancy. In Sudan seeds of Dalbergia sissoo stored less well at room temperature than those of local Acacia, Albizzia and Tamarindus species, while in Australia seeds of Acacia harpophylla deteriorate rapidly unless stored in sealed containers at 2–4 °C.

ORTHODOX SEEDS WITHOUT HARD SEEDCOATS

Many species in important genera of forest trees fall into this group, *e.g.* Pinus, Picea, Eucalyptus. Experience in Australia is that mature seeds of all eucalypts can be kept viable for some years if stored with a low moisture content in sealed containers at 3°–5°C. The majority of species can be stored for 10 years at room temperature with relatively little loss of viability. Both E. deglupta and E. microtheca seeds deteriorate more rapidly if stored at room temperature, but storage life is improved if they are kept in air-tight containers at 3°–5°C and recent evidence suggests that storage at-18°C is even better.

In Thailand seeds of P. kesiya and P. merkusii retained good viability for four years if stored at below 8 per cent moisture content in sealed containers at 0°–5°C, while at least five years' good viability is possible withP. caribaea and P. oocarpa under similar conditions. Considerably longer periods have been recorded for some species of pine *e.g.* 30 years for Pinus resinosa in USA when stored in sealed containers at 1.1°–2.2°C. Tectona grandis is an orthodox tropical broadleaved species but, since it produces good seed crops in most years, there has been little stimulus to investigate optimum conditions for long-term storage.

According to evidence summarized by Bowen and Whitmore, most Agathis spp. are orthodox. For example, an appropriate treatment ofA. australis in one study (dried to 6 per cent MC, then stored in sealed containers at 5°C) preserved viability for 6 years (79 per cent germination compared with the initial 88 per cent), while storage at below freezing temperature maintained a germination of about 60 per cent for up to 12 years. The same seed stored at higher MC or temperatures (15–20 per cent MC. or 15°–20°C) had lost all germination power within 14 months. A australisseed is inherently longer lived than A. robusta which in turn is longer lived than A. macrophylla. Initial trials with the tropical A. macrophyllaindicated that good results could be obtained by drying fresh seeds from about 65 per cent to 20 per cent before despatch by air (period in transit 14 days) and by further drying in the recipient country for 5 days at

16°C and 14 per cent RH. final MC was 6 per cent and germination 75 per cent. However, later trials were inconsistent and less successful. With tropical species, it is likely that handling between collection and despatch and the largely uncontrollable conditions of air transit are more critical than in the easier temperate or subtropical species.

Orthodox species which rapidly lose viability unless they are given the optimum treatment include species in the mainly temperate generaPopulus, Salix and Ulmus. Many of these lose viability within a few weeks under natural conditions or if stored in ambient conditions of temperature and humidity, but can be stored for months or years if maintained at low temperature and low moisture content. Examples are ofUlmus americana stored successfully for 15 years at 3 per cent moisture content and-4°C, and Populus sieboldii stored for 6 years at-15°C over a desiccating agent in a sealed container. In Populus balsamifera and Salix glauca, reduction in germination after two years of sealed storage at-10°C was less than 6.5 per cent of initial germination; after three years there was very little change in Populus but up to 40 per cent reduction in Salix.

In the tropics Aucoumea klaineana is a good example of an orthodox species which is short-lived under ambient conditions. Germination of fresh seed is often over 90 per cent, but after 30 days' storage in room conditions there is a significant drop in germination and this falls to zero after 100 days. Storage at 0–5°C and 7–8 per cent MC in sealed containers with a chemical desiccator, Actigel, maintains over 50 per cent germination for at least 30 months. There is some indication that a further reduction of MC will conserve viability even better.

Thus one seed lot of initial germination 76 per cent in the laboratory and 79 per cent in sand, when stored in sealed containers with Actigel, had an MC of 4.6 per cent and a germination of 70 per cent in the laboratory and 79 per cent in sand after 30 months; the same seed lot stored in sealed containers without Actigel had an MC of 9.9–10.4 per cent and a germination of 54–63 per cent in the laboratory and 62–67 per cent in sand. Other species of this type include Entandrophragma angolensewhich has a seed life of 6 weeks in room conditions but up to 6 years in cold storage and Cedrela odorata which loses all germination capacity in 10 months at room temperature but suffers no loss in 14 months if stored at 5°C in a sealed jar.

Some species may need special treatment to prolong viability for more than a few months. Fagus sylvatica can be conserved overwinter by maintaining MC at 20–30 per cent and storing part-filled in sealed polythene bags at 0–5°C for 100 days. It is then suitable for sowing because such storage conditions constitute a suitable pretreatment to break dormancy. If longer storage is intended, MC should be reduced to 8–10 per cent by drying in a current of air at room temperature (15–20°C). The nuts are then placed in sealed containers and stored at-5° to-10°C and will keep for several years.

Later research in France and Poland confirmed the above MC of 8–10 per cent and the advantages of sealed containers for long term storage. This technique has been successfully applied on a large scale (17 tons of beechnuts from 51 different sources) in France. Germination has been maintained over periods of 4 to 6 years.

Where storage conditions leave much to be desired, the longevity of orthodox seeds without hard coats can be expected to be much inferior to the hard-coated species. The nearer that conditions of storage approach the ideal for a given non-hardseeded species, the less the difference between its longevity and that of a hardseeded species. The best combination of MC and temperature will vary to some extent between species, for example the above-quoted 8–10 per cent MC for Fagus sylvatica is considerably higher than the 5–6 per cent considered ideal for many forest and agricultural seeds.

RECALCITRANT SEEDS

Recalcitrant seeds include a number of large seeds that cannot withstand appreciable drying without injury; it is of interest that the overwhelming majority of recalcitrant species listed by King and Roberts are woody. Temperate species such as Quercus andCastanea are commonly stored moist only for short periods over winter. Reduction of storage temperature to near freezing will prolong longevity.

Bonner found that it was possible to store acorns of Quercus falcata for 30 months and still obtain over 90 per cent germination at the end of the period, provided that temperature was maintained at 3°C and MC between 33 per cent (initial) and 37 per cent (final). A lower MC or a higher temperature (8°C) both reduced germination. For Quercus robur MC should be maintained above 40 per cent. Recent research in Poland has demonstrated good results from storing this species at >40 per cent MC in air-dry peat or air-dry sawdust in milk cans at-1°C. It is important to allow free entry of oxygen and this is ensured by inserting several strips of cardboard at intervals between the lid and the edge of the can.

In these conditions germination after 3 winters was in the range of 38–75 per cent and after 5 winters was still about 12 per cent. Temperatures below-5°C killed all the acorns, while a temperature of +1°C encouraged excessive pregermination (60–75 per cent after 3 winters, with radicles up to 25 cm long, compared with 12 per cent and radicles <0.5 cm long at-1°C). There may be possibilities of storing seeds after emergence of radicles. Recent research in Poland has indicated that best results are obtained with the recalcitrant Acer saccharinum by maintaining MC at the same percentage (50–52 per cent) as when the seeds were freshly collected. For A. pseudoplatanus in the UK a minimum MC of 35 per cent is recommended, while in Poland an MC of 24–32 per cent and a temperature of-3°C have proved suitable to store samaras over three winters.

Most short-lived recalcitrant tropical species are constituents of the moist tropical forests, where conditions conducive to immediate germination (high humidity and high temperature) are prevalent throughout the year. Typical genera are Hevea, Swietenia, Terminalia andTriplochiton, as well as a number of Dipterocarp genera such as Dryabalanops, Dipterocarpus and Shorea and some species of Araucaria.Dryabalanops is injured if dried below 35 per cent moisture content (MC) but still survives only about three weeks at over 35 per cent MC. Triplochiton seed is naturally short-lived but can be stored for up to 22 months at a temperature of around 6° C and a moisture content of between 12 and 25 per cent. Azadirachta indica seeds also have a short period of viability, although the species occurs in dry, not moist, tropical forests and it is not clear whether it is a genuine recalcitrant or simply a short-lived orthodox species.

Orthodox and recalcitrant species sometimes occur within the same genus. In Acer and Ulmus, genera in which both orthodox and recalcitrant seed behaviour occur, the distinction in North American species is clearly between spring-and fallseeders. A. rubrum and A. saccharinumflower and seed in the spring. Their seeds are not dormant, and their storage behaviour is clearly recalcitrant. Other Acer species have fall-maturing seeds, which are dormant and orthodox in nature at maturity.

The same occurs in Ulmus. Seeds of U. crassifolia and U. serotinamature in the fall, and are orthodox in storage behaviour. Spring-seeding species of Ulmus are 'weakly' recalcitrant. InAraucaria, A. cunninghamii and other spp. in the Eutacta taxonomic group behave as orthodox. In Queensland seeds of A. cunninghamii of 5 provenances were air dried and stored at varying temperatures in sealed and unsealed containers. At the higher temperatures, +1.7°C and-3.9°C, germination started to drop after 17 months' storage and after 8 years was down to about half the initial germination rate in sealed containers and about one third in unsealed containers. At the lower temperatures of-9.4°C and-15°C, germination after 8 years' storage was little changed from the initial (41–44 per cent compared with initial 49 per cent), and there was virtually no difference between sealed and unsealed containers.

The rate of viability loss at the higher storage temperatures varied from provenance to provenance, but all stored better at the lower temperatures. Moisture content was not recorded but under local conditions air-dry seed is normally in the range of 16–23 per cent. Later trials with Papua New Guinea A. cunninghamii have shown that seeds can be dried from 21 per cent to 7 per cent MC without any effect on initial germination rate; effects on storage life are still under investigation. A. hunsteinii in the Intermedia group and A. angustifolia, A. araucana and A. bidwillii in the Colymbea group are apparently recalcitrant. Arentz found that high viability of A. hunsteinii could be maintained for at least 6 months by storage at 3.5°C and high MC; 37 per cent was significantly better than 32 per cent. Research reported by Tompsett confirmed that MC should be maintained above 32 per cent. Placing the seed in one polythene bag of 25 microns thickness inside a second bag is effective in maintaining viability. The double thickness of polythene maintains a high MC but allows for some exchange of oxygen which is necessary to preserve viability of A. hunsteinii. A. angustifolia also needs a high MC; seeds died if dried to less than 25–30 per cent.

For some temperate recalcitrant species, as indicated above, a relatively low temperature (just above or just below 0°C) has been found beneficial in extending the life of the seeds; low temperature to some extent compensates for the high MC which must be maintained to prevent the early loss of viability. In some tropical species, seeds are quickly killed if temperature is reduced too low, just as they are quickly killed if MC is reduced too low. Among woody species cited in King and Roberts are Theobroma cacao (killed below +10°C), Mangifera indica(damaged below +3° to +6°C) and, among the dipterocarps, Hopea helferi, Hopea odorata and Shorea ovalis (damaged below, respectively, +5°C, +10°C and +15°C).

This susceptibility to chilling damage at temperatures above 0°C compounds the difficulty of storing these recalcitrant species, which seldom maintain viability for more than a few weeks or at most months. This compares with a normal seeding periodicity of several years in most dipterocarps, so there is no possibility as yet of conserving seeds in a viable condition from one good seed year to the next.

Unlike orthodox species, in which viability is best preserved by maintaining a minimal respiration rate, it appears that active respiration is necessary to survival of seeds of most recalcitrants. Thus damage to recalcitrant seeds has been reported not only from inadequate MC and too low a temperature but also from lack of oxygen *e.g.* in Araucaria hunsteinii, Hevea brasiliensis and Quercus spp.

Whereas some temperate recalcitrant species have been stored successfully for several years, seed longevity in tropical recalcitrants can be measured in days or weeks. The amount of research on tropical species is still small, especially on forest species, and it is possible that seed longevity could be prolonged beyond a few weeks if the best combination of seed maturity, speed, conditions and degree of drying, and most suitable storage temperature could be determined for each species. King and Roberts suggest a research strategy.

SEED COLLECTION, PROCESSING AND STORAGE METHODS

The benefits of exemplary seed collection, processing and storage methods may be largely lost if care is not taken over shipment from seed store to nursery. It is seed viability at the time of sowing, rather than at the time of despatch from the seed store, which determines the number of healthy plants produced from a particular seed lot. It is therefore essential to provide shipment methods which will ensure the minimum loss of viability in the interval between storage and sowing. The selection of appropriate packing material will depend on the characteristics of the species, the quantity to be shipped, the length of time in transit, the mode of transport and the temperature and moisture conditions to which the shipment will be exposel.

High and fluctuating temperatures and adverse humidity are the chief causes of viability losses during shipment. These factors are identical with those that cause deterioration in freshly collected fruits between the collecting site and the processing depot, as described. However, seeds between storage and sowing should start with advantage of having had optimum conditions of temperature and moisture content during the storage period. In fact, maintenance of storage conditions during transit would be ideal, but is often not possible.

Provided that the initial moisture content of the seeds is correct, it can be easily maintained during transit by the use of sealed containers. In some cases the seeds can be despatched in the same containers in which they were stored. In others it may be advisable to transfer them from a large container in storage to a smaller container for despatch. Individual nurseries may require only a small quantity of a given seed lot. In addition, small and light packages are often less subject to mechanical damage in transit than large, heavy ones. Magini recommends separate packages of 1–20 kg but not larger. A variety of moisture-proof or moisture-resistant material is available, as described earlier in this chapter under storage containers. Polyethylene of 4–8 mil (100–200 microns) has the advantage of restricting moisture passage while allowing exchange of oxygen and CO_2.

Sealed containers are highly suitable for orthodox species, of which the seeds must be kept dry during transit. The addition of a desiccant such as silica gel may be a useful additional insurance if there is any risk that the seeds may absorb moisture while being transferred from storage container to shipment container. Seeds of recalcitrant species, on the other hand, are best left unsealed, since the effect of some loss of moisture is less harmful than that of the overheating which can occur as a result of rapid respiration in sealed bags at ambient temperatures. They should be well-mixed with pulverized sphagnum moss, ground peat, coconut fibre or sawdust, that has been moistened and squeezed dry. A mixture of equal weights of dry packing and water will give adequate moisture content to these materials. In the case of international transit, however, an inert non-organic substance such as moist vermiculite is likely to be more acceptable to quarantine authorities.

Sealed moisture-proof containers should always be used for long journeys, *e.g.* form one country to another, of orthodox species of short longevity, provided that the initial MC is correct. But if orthodox seeds are being forwarded soon after collection and without having been dried to the appropriate MC for storage, it is preferable to ship in bags permeable to air rather than to seal with excessively high MC. A number of species with resistant seedcoats or pericarps, such as Tectona and many leguminous species, are able to withstand prolonged periods in ambient conditions; cotton or paper bags or hessian sacks are perfectly suitable for these species.

Large, moist seeds can be sealed individually with paraffin wax or latex. In the method described by Baldwin paraffin wax is heated to 71°–77° C and seeds or nuts dipped for a few seconds in a screen-type container, which should be shaken vigourously during the immersion. The waxed seeds should be packed in soft material so that the wax is not scraped off during transit. At the time of sowing the wax must be partly scraped off to permit the entry of water.

Protection against high or rapidly fluctuating temperatures is more difficult, but care should be taken to avoid placing the seeds close to local hotspots such as radiators and hot pipes. For very sensitive seeds, temperature effects can be mitigated by the use of insulating material in the packaging. Sub-zero temperatures do not usually affect dry seeds but may cause damage to recalcitrant seeds which must be kept moist. Premature germination is another risk which affects moist seeds. During storage, germination can be restricted by the use of low temperatures just above freezing, but the higher temperatures encountered during transit may induce germination in a substantial number of seeds. Seeds which are prone to germinate when held in moist packing may be treated with an inhibitor such as maleic hydrazide.

No matter what type of seeds is being despatched, it is necessary to take precautions against mechanical damage to seeds and against losses due to damage to the containers in transit. Double wrapping is often advisable, for example a sealed polythene bag should be placed inside a stout canvas bag. Stout drum cartons with sealed polythene or aluminium-foil containers inside provide an especially effective combination for seeds which need to be kept dry. If the inner bag is labelled, this is also an insurance against accidental defacement of the outer label. Clear labelling is essential and the consignee should be advised of despatch by means of an appropriate seed consignment note or seed issue form.

Stein et al. have provided a useful check list of helpful practices in seed shipment, reproduced hereunder:

- Double wrap the seed. Enclose the seed container in a sturdy, preferably rigid, outer container;
- Small or moderate size containers generally withstand shipment better than large containers;
- Fill containers completely to minimize air content and jostling of seeds during shipment;
- All packages should bear a good identifying label on the innermost covering and another one within the container;
- For long distances, shipment of sensitive seeds by air is desirable;

- Seed packages should permit ready opening and reclosing if destined for export to a country requiring fumigation. In addition a copy of the phytosanitary certificate should be readily available to quarantine authorities *e.g.* by sealing it in an envelope which is firmly attached to the outside of the package.

Seed storage facilities at nursery sites or district forest stations are inferior to those at the central seed store. Shipments should therefore be timed so that seeds can be sown with the minimum delay after receipt.

USE OF STORAGE SPACE

Some form of container is necessary for most seed storage, to facilitate access to, and handling of, individual seed lots while keeping them separate, to make the best possible use of storage space, to provide protection against animal and insect pests and, for some seeds, to prevent passage of moisture and gases between the enclosed and the outside atmosphere. Many types of container have been used for tree seeds.

They may be conveniently divided into:

- Materials freely permeable to moisture and gases
- Materials completely impermeable, when sealed, to moisture and gases
- Materials resistant, but not completely impermeable, to moisture.

STORAGE CAPACITY

The weight of seeds to be kept in store can be estimated in the manner indicated and will depend on the annual planting area, the maximum number of years' seed supply to be stored at any one time because of seeding periodicity, and the number of seeds per kg, for each species. Weight of seed in kg can be converted to net volume in litres (or in g to cm^3) by a factor related to average specific gravity. An average factor of 2.0 is appropriate for many forest species and corresponds to an apparent specific gravity of 0.5 (true specific gravity would be slightly higher because of the air-spaces between the seeds).

For conversion from net volume to gross storage space, allowing for shelving, ventilation, air spaces within and between containers, access and fittings within the cold room, a factor of about X8 is commonly used; this is with fixed shelving. Use of mobile shelving may double the quantity of seed which may be stored in a given space; in this case a factor of about X4 is appropriate. Thus 500 kg of seed of S.G. 0.5 would need a gross storage space of $500 \times 2 \times 8 = 8000$ litres or 8 m^3 if fixed shelving were used and 4 m^3 with mobile shelving. Where relatively few seed lots and large quantities of each lot are being stored, it is possible to use standard sizes of container, each filled to the brim, and for shelving space to be adapted to fit container size exactly. Under these conditions considerable savings in storage space can be effected.

Thus in the Danish seed store at Humlebaek, which uses fixed shelving, a factor of only 3.12 has been calculated.

DESIGN AND EQUIPMENT

The design and machinery for refrigerated storage is a matter for refrigeration engineers. Some guidance as to the features which should be included in any quotation for installation may be obtained from the excerpts from the IBPGR report which appears the example of the facilities installed by the Regional Genetic Resources Project at Turrialba which appears. It should be noted that both these documents refer to long-term storage of agricultural seeds for purposes of genetic conservation.

It is essential that designs and equipment be adapted to local conditions and local resources. The best installation in the world is of little use if it cannot be maintained, so it is essential to investigate the local provision for servicing and spares before committing oneself to any particular item. The reliability of mains electricity services and the need for a voltage controller and standby generator are of primary importance. Ready availability of a spare compressor may also be necessary.

The correct siting of a seed store may reduce the need for much expensive equipment. For example a tropical country with variable climate and topography might solve many problems by moving its store from a hot humid coastal site to the dry rain-shadow side of a mountain at 2000 m. In such a case a well-ventilated room might provide perfectly suitable conditions for several years' storage for relatively 'easy' species such as pines and eucalypts and could be supplemented by one or more deep-freeze chests for small quantities of more 'difficult' species requiring sub-freezing temperature. The value of deep-freeze chests was stressed by the IBPGR and its comments are reproduced.

MATERIALS FREELY PERMEABLE TO MOISTURE AND GASES

These include hessian or burlap sacks, cotton bags and containers of paper, cardboard and fibreboard. Hessian and cotton have the advantage that seed triers can be inserted through the cloth mesh to withdraw samples for testing without the need to open the mouth of the container. The resilience of the cloth will close the hole and avoid subsequent loss of seed, which is not possible with containers based on paper or paperboard Hessian and cotton are also robust materials which can be used more than once.

None of these materials is entirely proof against insect and rodent pests, and all are freely permeable to water vapour and gases. For orthodox seeds in uncontrolled conditions they are therefore suitable only for rather short storage periods; these can be extended in the case of hardcoated seeds or where ambient conditions are cool and dry. If seeds are stored in large containers after drying to the correct MC, the outer seeds themselves provide some barrier to the passage of moisture. Viability of the inner seeds may thus be preserved for a period even though there is some deterioration from increased MC in the outer layers. If a seed store has facilities for controlling both temperature and relative humidity, then permeable containers can be safely used for orthodox seeds for several years, provided that pests can be excluded.

For moist storage of recalcitrant seeds, open or freely permeable containers such as hessian sacks should be used in order to allow free exchange of air and so avoid the overheating which can occur if moist, rapidly respiring seeds are enclosed without adequate ventilation. Periodic spraying of the sacks may be necessary to maintain the high MC which is appropriate for this type of seeds.

MATERIALS COMPLETELY IMPERMEABLE WHEN SEALED, TO MOISTURE AND GASES

After drying of orthodox seeds to the correct MC, the MC may be maintained in storage by dehumidifying the whole storage space. Another very efficient way, commonly used in storing forest seeds, is to place the seed in sealed moistureproof containers. This avoids the need for expensive dehumidification equipment. For long-term storage the most effective method is a combination of moisture proof containers with controlled low temperatures provided by refrigeration.

An added advantage of most materials in this type is that they also exclude oxygen and so reduce still further the rate of respiration. Impermeable sealed containers are not suitable for storing recalcitrant seeds nor are they suitable for orthodox seeds at high MC, which deteriorate more rapidly in sealed than in open storage. Some seeds absorb moisture quickly, so it is important that they be sealed inside the container as soon as possible after drying is complete, preferably within the drying room itself.

Moisture proof containers include tin or aluminium cans and drums, glass jars of the Mason or Kilner types, plastic vials and laminated aluminium foil packages. Rigid and unbreakable metal cans provide maximum protection against mechanical damage to the seeds and are equally suitable for storage and subsequent shipment. Containers are only as moisture proof as their sealing. For rigid containers screw-top or clamp-down gasketed lids should be used, if periodic opening for seed extraction and subsequent resealing are anticipated; aluminium foil should be heat-sealed.

The effectiveness of sealing is particularly important in long-term storage. Three types of container are considered suitable for hermetically sealed long-term storage of agricultural seeds: glass jars or vials; metal cans; and laminated foil packets. They should be equally appropriate for orthodox forest seeds. But the report by IBPGR recommended sealed metal cans as the most reliable and convenient. It noted that the seals on screw-cap jars are not always perfect and that further experience of the lasting qualities of laminated foil packets is needed before they can be recommended for general use in storage which will often last for several decades.

MATERIALS RESISTANT, BUT NOT COMPLETELY IMPERMEABLE, TO MOISTURE

They include polyethylene and other plastic films and aluminium foil. These materials are resistant to the passage of moisture but, over a long period of time, there will be a slow passage of water vapour tending to equilibrate the RH inside with that outside the container. Some of the figures quoted by Justice and Bass for transmittal of water vapour appear surprisingly high, *e.g.* 0.13 g per 100 square inches (645 cm^2) per 24 hours for low density polyethylene film 10 mil (250 microns) thick and about ten times this figure for low density film 1 mil (25 microns) thick.

However, the standard conditions for testing these materials are 0 per cent RH on one side and 90–100 per cent on the other. The RH gradient during storage is never so severe as this and hence the rate of passage of water vapour is much less rapid in practice. In one test using 6 mil (150 microns) high density polyethylene the rate of passage over two years from an outside RH of 95–100 per cent at 20°/30°C was four times that from an outside RH of 50 per cent at 10°C. The thicker the film, the greater the resistance to passage of water vapour and, for a given thickness, high density polyethylene is more resistant than low density.

Although polyethylene is not suitable for long-term storage of orthodox seeds for genetic conservation, it is very suitable for short-or medium-term storage and has given excellent results for up to 5 years' storage of Pinus caribaea and P. oocarpa seeds in Honduras, with no significant change in MC.

For Honduran conditions a thickness of at least 4–5 mil (100–125 microns) is recommended; thinner polythene can permit a significant passage of water vapour in time and is also subject to mechanical damage in handling. Harrington considered 3 mil (75 microns) high density or 5 mil (125 microns) regular suitable for temperate conditions and 7 mil (175 microns) high density or 10 mil (250 microns) regular as adequate for even severe tropical conditions. Proper sealing of bags is essential and can be done by a combination of heat and pressure. In the past hot irons were used, but sealing can now be done more efficiently and conveniently by commercial heat sealers, of which a number of different models is now on the market.

Different materials, each alone slowly permeable to water vapour, may become completely impermeable when laminated together. Various combinations of laminated polyethylene, aluminium foil and kraft paper proved completely impermeable to water vapour over a two year period, even when there was a high differential between the inside and outside RH.

USE OF DESICCANTS IN CONTAINERS

If orthodox seeds are dried to the correct MC and stored in sealed impermeable containers, the MC should remain constant for years. If, however, the seeds are stored in moisture resistant but not completely impermeable material such as polythene bags, or if it is necessary to open and reclose the containers periodically to extract seeds, there will be a slow build-up of moisture in time. A convenient way to prevent this is to enclose some desiccant such as silica gel in the containers. The capacity of silica gel to adsorb moisture depends on the relative humidity of the ambient air, as shown in the following table:

Table. Moisture Content of Silica gel in Equilibrium With Various Relative Humidities

% RH	% H_2O Adsorbed	% RH	% H_2O Adsorbed
0	0.0	55	31.5
5	2.5	60	33.0
10	5.0	65	34.0
15	7.5	70	35.0
20	10.0	75	36.0
25	12.5	80	37.0
30	15.0	85	38.0
35	18.0	90	39.0
40	22.0	95	39.5
45	26.0	100	40.0
50	29.0		

A convenient method is to use silica gel treated with cobalt chloride, which changes colour from blue to pink about 45 per cent RH; the corresponding equilibrium MC for many orthodox species would be 7–9 per cent. Dried silica gel is enclosed with the seeds and, whenever the granules turn pink, the silica gel is removed and reactivated by drying in an oven at 175°C and cooling in a sealed container before reuse. A weight of silica gel equal to one tenth the weight of seeds is recommended. Care should be taken not to include too much silica gel which could lead to overdrying of the seeds. Even with silica gel of one tenth the weight of seeds, the MC of seeds enclosed at 6 per cent would be lowered to below 5 per cent during the initial phase of storage. More frequent reactivation of silica gel would preserve an equilibrium of RH and seed MC at lower levels than the 45 per cent and 7–9 per cent mentioned above but would forego the convenience of the colour indicator.

1 kg (oven-dry weight) of seed at initial 19% MC (dry weight basis) contains		190 g H_2O
1 kg (oven-dry weight) of seed at 6% MC (dry weight basis) contains		60 g H_2O
Therefore moisture to be removed	=	130 g H_2O
RH in equilibrium with 6% MC	=	25 % RH
At RH 25%, 1 kg silica gel adsorbs		125 g H_2O

Another use for desiccants is where the MC of seeds is known to be higher than the optimum for sealed storage, for example because only air-drying is possible. As mentioned, the enclosure with the seed of approximately an equal weight of silica gel in sealed containers should reduce the MC of the seed to a suitable level and maintain it. As an example, Therefore a weight of silica gel equal to the weight of seed will reduce the initial 19 per cent MC to just over 6 per cent MC for storage.

CHOICE AND USE OF CONTAINER

The following factors, which should be considered when choosing the best storage container for a given use, are based on those listed by Stein *et al.*: When seed requires further drying in storage, do not use a tight-closing container because enclosing excess moisture is harmful to the seed. Use a tight-closing container if gain in seed moisture content can be damaging and relative humidity in the storage facility is high. Containers and seed can quickly gather unwanted condensation when brought out of cool or subfreezing storage. Warming to room temperature is recommended before opening a container brought out of such storage. 4 to 10 mil (100–250 microns) polyethylene bags will greatly restrict exchange of moisture, but still allow exchange of oxygen and carbon dioxide with air outside. Such exchange may be beneficial or harmful, depending on species.

A container that is easy to open and close is desirable when quantities of seed are likely to be added or removed repeatedly. In order to minimize temperature and relative humidity fluctuations, open only when necessary. Alternatively, store seed in small containers, so that the entire contents can be stored or emptied at one time. For orthodox seeds, fill containers completely to ensure minimum exchange of moisture between the seed and the entrapped air and, more importantly, to limit the amount of oxygen enclosed. When exchange of moisture through the container walls must be eliminated or restricted, the container must be made impermeable or of moisture resistant material. The longer the storage period and the higher the differential between external RH and RH within the container, the more impermeable the material must be.

ARTIFICIAL SEED AGEING

Vigour testing by ageing tests was elaborated by Delouche and Baskin for estimating the storage potential of seed lots and is now accepted for all species with orthodox storage behaviour. Later, ageing tests were also shown to indicate the relative field emergence of seed lots of numerous species. The principle of ageing tests is to expose seed samples for a defined period of time to an unfavourable environment of high temperature and high seed moisture content. After this ageing period, high vigour seeds are expected to still show high germination, whereas low vigour seeds are expected to show a considerable decrease in germination. The controlled deterioration test, known as CDT is presumed to mimic natural seed ageing. CDT is widely used as a vigour assay for numerous seed species and has been described. We have used this protocol to deteriorate in a controlled manner Arabidopsis or sugarbeet seeds. Seeds are equilibrated at 85% relative humidity (20 °C), and day 0 controls are immediately dried back at 32% relative humidity. Treatment is done by storing the seeds (at 85% relative humidity) for 7 d at 40 C. Then these seeds are also dried back at 32% relative humidity (20 °C) during 3 d.

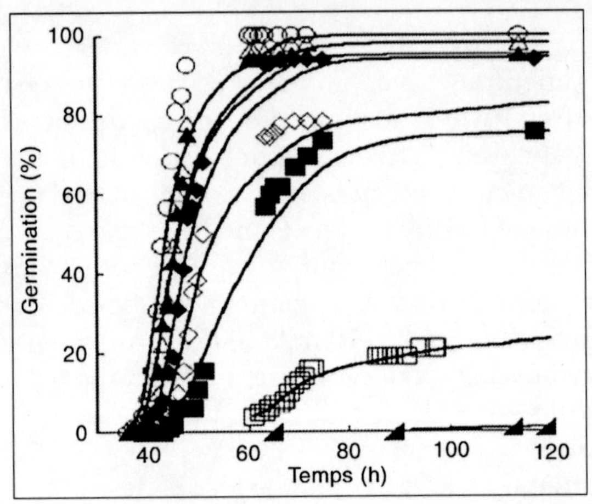

Fig. Germination Curves for Seeds

AGEING TREATMENTS

Experiments were carried out with sunflower seeds harvested in 2004 and received from Monsanto-France. Seeds were stored for 3 months at 20 °C and 75% RH in order to break their dormancy before ageing treatment. Ageing was performed according to Kibinza *et al.* Seeds were equilibrated for 24 h at 20 °C, in closed flasks with different amounts of water to obtain seed MCs ranging from 0.04 g H_2O g^{-1} dry matter (DM) (dry seeds) to 0.48 g H_2O g^{-1} DM, and then placed at 35 °C for 7 d. Flasks were regularly flushed with air to ensure seed oxygenation.

After ageing treatment, germination assays were performed on whole seeds (whole embryo, *i.e.* embryonic axis plus two cotyledons, plus the seed coat) and embryonic axes (radicle plus gemmula) were isolated and frozen for DNA extraction or used fresh for adenosine phosphate assay and cytological experiments.

GERMINATION ASSAYS

Germination assays were performed at 15 °C in darkness, in three replicates of 50 whole seeds placed in 9 cm diameter Petri dishes on a layer of cotton wool moistened with deionized water. Germination counts were made carried out for 7 d. The results presented correspond to the means of the germination percentages obtained after 7 d ±SD.

WATER CONTENT DETERMINATION

Water content determination was determined on 10 whole seeds according to Kibinza *et al.*

ADENOSINE PHOSPHATE MEASUREMENT

The adenylate pool (AP: ATP, ADP, and AMP) was extracted according to Olempska-Beer and Bautz-Freeze from three axes excised from seeds after 1 d of water equilibrium at different MCs at 20 °C and after a subsequent 7 d controlled deterioration at 35 °C. The components of the AP were measured using the bioluminescence method with a pico-ATP biophotometer, as described by Corbineau *et al.* and Kibinza *et al.* Values were calculated as nmol AP per mg DM and are the means of values obtained with 5–7 extracts. Data were subjected to an analysis of variance using Duncan's multiple range test at P d 0.05.

DNA EXTRACTION AND ANALYSIS

Embryonic axes isolated from whole seeds using a sharp scalpel blade were immediately frozen in liquid nitrogen, and then stored at –80°C. DNA was isolated from embryonic axes according to the method described by Goldenberger *et al.* A 0.2 g aliquot of sample was ground with liquid nitrogen by using a mortar and pestle for ~ 5 min. The resultant fine powder was suspended in 2 ml of SDS extraction buffer (1% SDS, 200 mM TRIS-HCl, pH 7.5, 288 mM NaCl, and 25 mM EDTA) by vigorous vortexing for 2 min. Debris was removed by centrifugation at 10000 g for 2 min. The supernatant was mixed with an equal volume of cold isopropanol and then the precipitate was separated by centrifugation (12000 g, 7 min). The nucleic acids in the pellet were dried and then dissolved in 100 ml of TE buffer (100 mM TRIS-HCl, pH 7.5, and 1 mM EDTA).

DNA electrophoresis was performed to assess DNA fragmentation. DNA samples (10 μg lane^{-1}) were loaded on a 1.5% agarose gel stained with 0.2 μg ml^{-1} ethidium bromide.

RAPD ANALYSIS

Amplifications were carried out in 25 μl of reaction mixture containing 100 ng of genomic DNA, 120 ng of primer, 200 μM dNTPs (50 μM of each), 10× reaction buffer, and 0.5 U of Taq DNA polymerase.

The RAPD protocol consisted of an initial denaturing step of 5 min at 94 °C, followed by 40 cycles at 94 °C for 1 min, 34 °C for 1 min, and 72 °C for 1 min, with an additional extension period of 10 min at 72 °C. RAPD products were analysed by electrophoresis in 1.5% agarose gels stained with 0.2 μg ml^{-1} ethidium bromide. The polymorphism presented is repeatable and corresponds to representative observations obtained with three different DNA preparations from independent biological replicates

EFFECT OF AGEING ON ENERGY METABOLISM AS RELATED TO SEED MOISTURE CONTENT

The AP (ATP, ADP, and AMP) was measured in the axis before (upon seed equilibrium with various amount of water) and after 7 d of ageing at 35 °C. The decrease in the AP, which represents the difference between the AP content before and after ageing, is expressed as a percentage of the AP measured in unaged seeds equilibrated at various MCs. This mode of calculation allows integration of the effect of increasing MCs on the AP. Seven days of controlled deterioration at 35 °C resulted in a decrease in the AP that was related to seed MC but not proportionally, thus allowing description of two levels of mitochondrial impairment. From 0.14 to 0.29 g H_2O g^{-1} DM, the AP decreased by 25% after ageing. When the seed MC was >0.37 g H_2O g^{-1} DM, the pool decreased markedly and represented 40% of that found in unaged seeds.

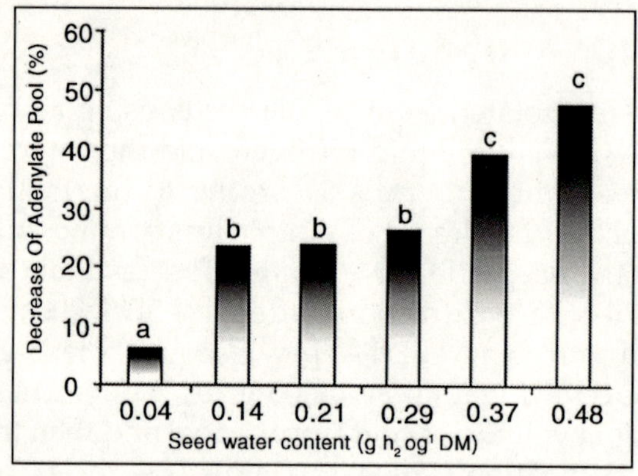

Fig. Effect of 7 d of Controlled Deterioration at 35 °C of Seeds Equilibrated at a Water Content of 0.04, 0.14, 0.21, 0.29, 0.37, and 0.48 g H_2O g^{-1} DM on the Secrease of the Adenylate Pool (ATP, ADP, and AMP).

RELATIONSHIP BETWEEN SEED MOISTURE CONTENT AND DNA ALTERATION

RAPD profiles obtained with DNA templates from seeds aged at various MCs are shown in Figure. Among the 20 decamer oligonucleotide primers tested, only 13 gave specific and reproducible results. Several primers (1, 2, 3, 6, 8, and 16) gave a constant pattern of PCR products in all DNA samples, whatever the seed MC during ageing. In contrast, other primers showed polymorphism and gave different PCR products as a function of seed MC. Primer 12 amplified two bands only in dry seeds (0.04 g H_2O g^{-1} DM) and did not gave PCR products at higher MCs. For primers 5, 7, 10, 11, 15, and 17, the greatest variability in the pattern of PCR products occurred when the seed MC was 0.37 or 0.48 g H_2O g^{-1} DM.

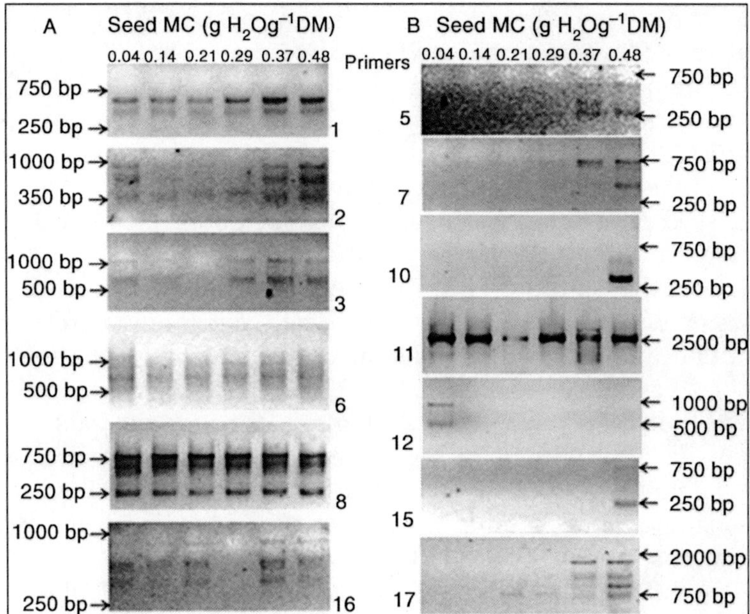

Fig. Representative Electrophoregrams of Amplification Products Obtained from the DNAs of the Axis of Aged Seeds Containing a Water Content of 0.04, 0.14, 0.21, 0.29, 0.37, and 0.48 g H_2O g^{-1} DM.

Primer 17 allowed a slight amplification when the seed MC was 0.21 g H_2O g^{-1} DM, but the number of bands increased to about three and four when the MC increased to 0.37 and 0.48 g H_2O g^{-1} DM. Primer 11 amplified five bands only with the DNA sample of seeds at an MC of 0.37 g H_2O g^{-1} DM. At the higher MCs, progressive and extensive DNA degradation probably induced a lack of sites for primer 11 but a gain of sites for primers 10 and 15. The general trend (*i.e.* six primers among the seven primers which give a variable pattern) was an increasing number and amount of products with increasing seed water content, and major changes in DNA occurred when the seed MC reached 0.37 g H_2O g^{-1} DM.

AFFECT OF AGEING

Seed ageing is influenced by two environmental factors, RH and temperature. Accelerated ageing has been developed under laboratory conditions by exposure of seed to high temperatures (30–45 °C) and high humidity. Accelerated ageing has been recognized as a good predictor of the storability of seed lots. It is associated with loss of vigour (*i.e.* the ability to germinate rapidly in a wide range of environmental conditions) and then with a progressive loss of viability. The loss of viability can be modulated by exposure of seeds at constant MC to different temperatures or by exposure of seeds equilibrated at different MCs to a constant temperature.

In fact, 7 d ageing of sunflower seeds at a constant MC (0.29 g H_2O g^{-1} DM) resulted in 90% germination when exposed at 25 °C, 80% at 35 °C, and 0% at 45 °C compared with non-aged seeds which displayed 98% germination. The present work shows that loss of seed viability during ageing at a constant temperature (35 °C) strongly depends on the MC, as was previously described with seeds of the same species. This behaviour allows a critical MC at which 50% of the seed population was dead (P50) to be defined. Under the present experimental conditions this value was 0.37 g H_2O g^{-1} DM.

It is shown that sunflower seed ageing is associated with increased DNA damage, as evidenced by RAPD analysis. These data are in agreement with previous publications which demonstrated that DNA deterioration is a key feature of seed ageing using either RAPD or AFLP techniques. Interestingly it is shown that the extent of DNA damage depends on seed MC, since polymorphism varies with increases in seed MC. This suggests that seed MC regulates the nature and/or the rate of DNA alteration, as shown by RAPD analysis. A single primer produced a pattern of disappearance of bands when seed MC increased above 0.04 g H_2O^{-1} DM.

DNA modification related to repair has been shown by Boubriak *et al.* in the early steps of seed imbibition, and this mechanism has been proposed as being a prerequisite for allowing the completion of the germination process. The most common kind of variability in the patterns of RAPD-PCR products nevertheless consisted of the appearance of new bands. This trend has already been reported during ageing of soybean seeds. New fragments can be amplified because some sites become accessible to the primer after structural changes in the DNA take place. This could be due to point mutations and/or extensive rearrangements of the DNA. A single point mutation within the primer site can generate significant changes in RAPD patterns. DNA alterations evidenced by polymorphism were observed when seed MC reached at least 0.37 g H_2O g^{-1} DM since lower MCs had no effect on the pattern of bands. Interestingly, this is also the MC at which DNA fragmentation showing the typical pattern of laddering was evidenced. This pattern is considered as being a hallmark of PCD, thus suggesting that this active mechanism of apoptosis is likely to take place during sunflower seed ageing under hydrated conditions.

The involvement of PCD in loss of viability was confirmed using TUNEL assay and by studying the changes in the AP. The strong depletion of the AP that occurs at 0.37 g H_2O g^{-1} DM suggests mitochondrial dysfunctioning at this MC which can result from changes in the mitochondrial ion channels and release of cytochrome c into the cytoplasm. A decrease in the AP was recorded from an MC of 0.14 g H_2O g^{-1} DM, which suggests an impairment in mitochondrial metabolism at this ageing point corresponding to 90% germination. It has been reported that the impairment of energy metabolism is a process which occurs when cell viability is almost unaffected, which suggests that mitochondrial damage occurs in the early phase of PCD in plant and mammalian cells.

Primary events of PCD are triggered by ROS, and a threshold level of ROS is required to activate the signal transduction pathway that results in PCD. It was shown using the same model and ageing conditions that ROS and lipid peroxidation increased according to ageing intensity, namely MC. Oxidative stress and lipid peroxidation have been widely indicated as the major cause of deterioration of oilseeds during ageing. Furthermore, changes in the half-cell reduction potential of glutathione, which represents the major cellular antioxidant and redox buffer, were related to PCD during ageing in pee seeds.

In situ observations provide supplementary useful data because they allowed PCD to be studied at the level of individuals, whereas DNA fragmentation and mitochondrial failure were evidenced at the population level. These observations show that the whole cellular territory of seeds undergoes PCD in a synchronous manner when they have reached a critical point resulting from a combination of temperature, duration, and MC applied during the ageing treatment (close to 35 °C, 7 d, and 0.37 g H_2O g^{-1} DM under the present conditions). If all cells simultaneously undergo PCD, as shown here, the seed will die and this mechanism cannot therefore be considered as a selective elimination of unwanted cells, as was shown in other developmental processes or in response to pathogens.

11

The Practice of Long Term Seed Preservation

An alternative sometimes used consists of controlling moisture in the cold room itself. It is not a bad idea itself, although it shows some disadvantages. With a temperature of about 0 °C (more than sufficient as it has just been explained), to reach a relative humidity of approximately 25% inside the chamber is neither difficult nor expensive. Anyway, this would be far more efficient than having a chamber at -20/–25°C with the containers tightness being neglected. However, controlling the moisture content within the containers themselves by using silica gel is, in the long term, much more practical and efficient, since it allows reaching more suitable ultradry levels and a better monitoring.

Ultradrying with silica gel can be achieved by putting the samples inside a closed container along with the dehydrated silica gel.

The use of flame-sealed vials, obtained from standard laboratory glass tubes of 20 ml is a system whose high efficiency has already been proven. If it cannot be generally recommended, is only on the grounds that it allows little volume – a maximum of 8 ml - for large accessions. However, it should be considered that 8 ml means some 160 seeds of wheat, 1400 of cabbage and 40000 of Diplotaxis siettiana (extinct in Nature). The method is almost perfect for especially valuable material such as rare, endemic or threatened species, special breeding material, "black-box" collections, etc. Also, the idea of preparing several vials for each accession is realistic and most often worthwhile.

In the following paragraphs, the different steps to create such ampoules according to the procedure used in the UPM genebank are explained in detail:

1. The test tubes must be made of alkaline glass. This can be asserted through some greenish colour in their rim. Other glass types might soften more slowly or can even crack when heated.
2. Since long term preservation is intended, it is wise to place inside an internal label with the accession number, because any external label could get deleted or lost after a time. The number is easily

written with a ballpoint pen on the sticky side of a label and then pressed against the wall with a rod.

3. The seeds, already clean and dry, will be introduced from another tube or by means of a funnel. Although it is possible to fill up to 8 ml, a less quantity is often placed in order to have more vials available when it becomes necessary to break one in the future. Rubber stoppers in Figure are intended to avoid provisionally moisture intake from the outside.
4. A Piece of cotton, preferably hydrophobic and previously dehydrated will be used to separate the seeds from the silica gel which will be placed on top.
5. Afterwards, approximately 3 ml of dehydrated silica gel will be added. Before moving on to the next step, it is advisable to place a rubber cap on the tube and wait for 15–20 days. If the seed is not properly dry, this will be shown in the bottom of the gel and simply changing this gel could be enough to take the sample to the desired moisture balance.
6. Another piece of cotton, like the one used in (4), will help to keep the whole in place. Around 5 cm of the tube should be left free, to allow sealing it by heat.
7. A numerous group of tubes can then be placed (convenient) for 24 hours within a box which is filled with CO_2. Any other inert gas could do the job but CO_2 being heavier that air may be managed more easily. Cylinders with dry CO_2 are easily obtained from the market.
8. The flame from a standard laboratory burner may not be enough for a comfortable sealing of the tube. Therefore, a second entrance hole will be useful to apply a stream of oxygen or air. The current supplied by a standard house hairdryer could be enough to stoke up the flame.
9. The sealing itself can be done reasonably quick once some practice has been acquired. Holding the tube by its base with the fingers (no heat is felt) and with a long tong on the upper rim, concentrating the flame action and turning to spread the heat around, approximately 100 vials per hour can be thus closed.
10. Immersion of the sealed vials in water for 24 hours, would allow to check if any of them was not properly closed—the silica gel will change colour. The procedure should then be repeated for this particular case.
11. It would only remain necessary to add an external label with more data, according to the preferences of each bank, and to protect somehow the more delicate part of this set, the flame sealed tip. Both goals can be combined. A long, sticky label is rolled around

the top and is used as a plank mould to pour inside a solidifying liquid.

12. Some substances were tried, such as melted wax, epoxy resins, etc. The conclusion reached was that the most practical solution consists of a mixture of wax (2/3) and pitch (= colophony or resin) (1/3)–This last substance used to harden the wax. The mixture is melted and poured with a dropper in the hole created by the label around the glass tip.
13. In order to avoid the top surface becoming scratched, a nail varnish could be applied.

When stretched with the flame and externally labelled, the vials should become of a final homogeneous size. In the UPM's genebank they have traditionally been stored inside Kilner jars—long before finding that the last also were tight enough. Around 25 vials can be placed in each jar—thus obtaining a double security.

The described protocol might seem difficult and work demanding but this is not at all the case when the work is properly set up and organised. The seeds can thus be preserved without any evident viability loss for at least 40 years , which may well mean an efficient preservation for at least one century or probably more. The advantage and the disadvantage at the same time, is that the vials can only be opened by breaking them, unlike what happens with the Kilner jars.

The Detection of sufficiently reliable larger containers as the Kilner jars opened the way for the adaptation of the silica gel method to the larger samples that are common in crop genebanks. In this case, three important differential aspects should be taken into account

Firstly, Drying in situ is trickier than it was with tubes. Since the seed sample may have an initial moisture content of approximately 8–12%, the gel will be hydrated by the sample and will change its colour. It will be necessary to replace this gel with a new, dehydrated one, and the operation might need to be repeated several times. Once the gel does not change colour, one more replacement is convenient. The seed will then be ultradesiccated and will remain indefinitely balanced with the dry gel if the container is kept tight. However, changing the gel repeatedly might be fastidious. Therefore, it is convenient to dry the seeds as much as possible by some other method (sun exposure, dry room, dry air currents or other drying agents such as Cl_2Ca). In other words, it is advisable to start ultradrying with the seeds already as dry as possible. Drying speed will depend of the mass ratio between the desiccating agent and the seeds but once the equilibrium is reached, only a small amount of silica gel to act as indicator is enough (and also convenient, see later).

Plastic chambers (not long-term tight, but tight enough to perform this operation, can hold a series of cardboard boxes, some with gel and other with seeds. By changing the gel when necessary, it will be possible to desiccate inside a high number of samples in only a few weeks. Since most seed collecting missions at least in the Mediterranean area take place in June or July, it is possible to use the period of August - September to desiccate the seeds.

If there are funds available to purchase new equipment, and once the efficiency of the method becomes well established, a previous freeze drying can take the seeds directly to moisture levels very close to the balance with dehydrated gel.

Ready-to-use desiccators containing silica gel which is automatically regenerated already exist in the market. They usually consider that the chamber where the samples would be placed will be provided by the user. When the chamber is a store room, even tiled, big or small, it is inevitable that humidity will come inside through the walls. This makes it difficult to achieve relative humidity levels below 15%, especially with low temperatures. Besides, the drier constitutes a dynamic system with several pieces, joints, etc., very far from the simplicity of the jar of Figure (where <5% can be reached). Compact systems with built-in chamber to be placed on top of a laboratory bench are already present in the market. However, these cabinets never guarantee a relative humidity under 20%. R.H. under 15% could perhaps be reached in the future with well designed impermeable glass or metal chambers. However, if the final goal is to store the seeds in Kilner (Scotch) jars, a previous storage at 15–20% R.H. in the cold room may help to reduce the number of times the silica gel must be replaced.

Secondly, it is important to take into account that silica gel can absorb important amounts of water (up to 20% of its own weight) and lose part of its efficiency before changes in colour can be appreciated. This is particularly important with jars that are frequently opened and closed. To obviate this problem fully dehydrated silica gel should always be used and, once the equilibrium seed/gel has been obtained, to place only a small amount of gel (*i.e.* a small bag in the upper part) is sufficient. In this way, any moisture intake will be detected soon.

In the third place, it is convenient to minimise the presence of oxygen, either by filling the containers as much as possible with the seed material or by replacing the air with an inert gas, preferably CO_2 because it is heavier and tend to occupy the lower parts of a volume as water do. A few open Kilner jars could be placed in an "ad hoc" box where CO_2 is added. Otherwise, CO_2 could be directly added from the cylinder if this is accompanied by a pressure reducing device. Also, unlike sealed vials, a Kilner jar may be opened, perhaps frequently, and this has to be taken into account by adding CO_2 once in a while. It should be noted that the substitution does not need to be perfect because the aim is to reduce (not necessarily eliminate) the O_2 present within the container.

The same system (Kilner jars with silica gel at the bottom) could be valid to stop aging in badly preserved seed collections. It would be enough to place the samples inside (in bulk, in their original containers or better in paper envelopes) and to regenerate the gel as many times as necessary until no further colour change would occur. An additional substitution of the gel will ensure that this is completely dehydrated. In the particular case of foil bags, several of them could be fitted inside each jar. Each jar would be afterwards given a number and the accessions position within the genebank then recorded in the database.

GENERAL PRINCIPLES OF SEED

Low moisture, low temperature, low ethylene concentration and probably low oxygen concentration are the most relevant factors to have in mind for an efficient long term seed preservation. Some interrelations among them exist.

Low moisture and low temperature have been the traditional factors considered. Still, in the practice of the past few decades, low temperature has comparatively received much more attention, while an efficient control of low moisture has been largely neglected. However, in order to guarantee an efficient long term preservation of orthodox seeds, a good control of seed moisture content is essential. At this respect, two major principles should be taken into account:

The container should be perfectly tight, preventing water vapour intake. If the container is not perfectly vapour tight, the seeds will tend to get balanced with the external air humidity. In this way, any potential benefit of low temperature will be offset by an increase in seed moisture. It must be noticed that relative humidity inside any uncontrolled cold room is usually very high. Further to this, it should be emphasised that dried seeds are highly hygroscopic and–even if often very slowly—there is time enough ahead for moisture to enter the container. Unfortunately, the utilisation of inadequate containers has been commonplace in seed genebanks (Gómez-Campo, 2002).

Ultradrying down to a moisture content of 1–3% might extend the seed life span perhaps 4–16 times (Harrington, 1972). Ultradrying has been considered harmful by some authors (Walter & Engels, 1998; Hu, C. & al. 1998), while others have defended that it is not damaging to orthodox seeds (Ellis, 1998; Hong & al. 2005).

The debate might last for years until the subtle factors behind these opposite opinions are fully understood. However, under the conditions utilised in the Universidad Politécnica de Madrid (UPM) genebank, ultradry seeds preserved during 40 years have maintained their viability intact (average = 98,4%, not significantly different from 100% ; Pérez-García & al., 2007). As far as we know, these results are unmatched by any other seed genebank and set a solid base to develop improved preservation methods. This has moved the author to prepare the present condensed guide.

When dealing with ultradry seeds, enormous savings of energy can be made by avoiding the use of too low temperatures during storage. Temperature may be the key factor when seeds are merely dried but it loses relative importance for ultradried seeds (Pérez-García & al., 2007). Independently, very low temperatures, close to that of liquid nitrogen (–196 ºC), provide an alternative method for the preservation of seeds and tissues (cryopreservation) that might play an important role in the future, especially for recalcitrant material.

A third factor—the presence or absence of oxygen–has been the source of a contradictory literature over the years and, perhaps for this reason, it has largely been neglected from the practical point of view. However, recent research by Ellis & Hong (2007) suggests that ultradry seeds may be sensitive to oxygen; therefore, it is a factor that, at least tentatively, should be taken into account. Perhaps, the success of the UPM bank partly resides in the fact that air was originally substituted by CO_2 in the atmosphere within the containers. Ethylene and other toxic gases that slowly evolve during seed aging should be removed in order to extend the seed life span.

CHOICE OF CONTAINERS

For many years it has been widely admitted that containers which are appropriate for cold drinks or to store chips for a few months, are also appropriate for long term seed storage. This is very far from reality. Results of a survey on 40 different containers—some of them widely used in many genebanks-showed that 36 of them (90%) allowed moisture inside in less than three years.

All plastic containers allowed moisture inside. The water molecule is small enough to get through the pores of the polymers used to make the plastics and in the medium and long term the fact is unavoidable–although not noticeable in the short term. Polyvinyl bags are especially permeable to humidity. An additional reason for the failure of containers with a lid is based on the fact that both pieces are usually made of different materials, therefore expanding or contracting in a different way with changes in temperature. This creates fissures allowing humidity to get inside the container. This last mechanism appears not only in plastic containers, but also in most glass containers with twist-off or screw lid. Even plasticized twist-off type jam jars often show oxidation stains inside after some time has elapsed.

A special mention should be made to aluminium foil bags because of their still widespread use in genebanks of cultivated species. It's been a long time since bi-laminated bags (coated with plastic only on one side) were rejected on the basis of their inefficiency. However, tri-laminated bags (coated on both sides) are still widely accepted.

According to our experience a maximum of 80% of a given set keeps vapour tight after 3–6 years. It is evident that to base the whole operation of a genebank on a material with these characteristics is risky. To put an example forwards, a collection of 100.000 samples where we learn that 20% of them (20.000) won't keep their tightness creates a really awful scenario which is now quite common in genebanks. Additionally, it is not possible to see from the outside in which ones the problem is. Massive germination tests or even measurements of moisture in all the samples, are not practical solutions. Let's add to it that, being sealed with plastic, they carry on the long term the disadvantages of this material.

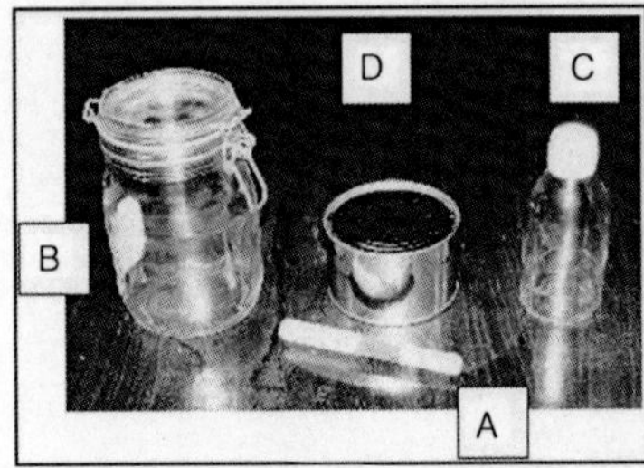

Fig. (A) Flame sealed vials (B) Kilner (Scotch) jars (C) Flask for laboratory chemical (D) Sealed metal cans

The aforementioned experiment showed that only four containers remained vapour tight:

a. Flame-sealed glass vials. Used in the UPM's genebank since 1966, they have only very exceptionally (0,3%) showed cases where moisture had come inside, due to the sealed tip having got broken. This tip can be protected with a mixture of resin and wax, as will later be explained. Undoubtedly this is the best option for small valuable accessions of wild or crop species such as rare or endangered species or genotypes, crosses, mutants, and so on.
b. "Kilner" ("Scotch") jars with a rubber joint and propped up lid, whose use is quite spread in kitchens. After 12 years, only one out of 25 units allowed moisture inside. Its size allows their use with samples of every size, including those bulky ones often kept in genebanks of cultivated species. A good adjustment of the joint should be ensured and also that nothing sticks between the glass and the joint (adhesive tape, little seeds or gel grains). It is possible that some joints need to be replaced after several years. As they are hermetic, seed samples can be stored inside in bulk, or in any type of smaller container such as paper or polyvinyl bags.
c. Some glass flasks with plastic lid, usually used in laboratories to store chemicals behaved very well during the experiment but other very similar ones behaved very badly. The difference was due to the quality of the plastic joint under the lid. Their opening being narrow, these jars are somehow uncomfortable to be used with medium sized or large seeds.

d. Heat sealed, metal cans coated with plastic on the inside. This inside plastic coating may be beneficial to prevent any possible damaging effects from the metal vapours in contact with the seeds over the years. The seal is also plastic. On the other hand, these containers are opaque and do not allow external control with coloured moisture indicators - unlike the three previous ones.

Those vials with a rubber lid fastened by an aluminium cover, used some years ago to contain penicillin, were not tested. However, users of this container refer to its behaviour positively.

It is very possible that other appropriate containers could be found in the future. However, to do reliably so, perhaps several years might be necessary. On the other hand, models very similar in appearance—case (C) and also some (B)—can behave very differently and it is advisable to test them before choosing a particular one. The method there described might serve to test other containers. Considering that most banks around the world have too often used non vapour-proof containers, this experiment probably shows the main reason of failure in seed preservation – especially for banks of cultivated species. It also shows some possible solutions. Solution (B), in particular, provides a larger volume and opens a way to the use of the silica gel method in seed genebanks of cultivated species–A method already used successfully since 1966 with wild species (Gómez-Campo, 1972).

On the other hand, it is obvious that the containers tightness is linked, in a practical way, to the need for frequent or only occasional handling of the seed lots. For very long term collections of valuable materials addressed to the future ("black-box" collections), there is no doubt that the flame-sealed glass vials are the best option. Those can also be of help for other material used more often, especially from wild species or any small sized species, by placing smaller quantities of seeds and making more vials, so that one of them can be broken when necessary. For base (long term) collections, Kilner jars are also advisable and so are the other options already mentioned. Active collections (for distribution), require more frequent access and need containers capable of being opened and closed. Kilner jars may also be used here if being fast when taking the sub-samples and, whenever possible, doing it in a dry environment.

For active collections of species with small seeds, tubes with a screwed plastic lid might be of interest. They are easy to use but not entirely reliable. Transparency of the containers becomes a key quality when wishing to take advantage of the benefits of the silica gel method for an easy and direct monitoring of the samples from the outside.

ULTRADRYING

To this date, the most commonly used procedure has consisted of drying the seeds with different procedures until reaching a moisture content of approximately 5–7%. After this, the seeds are placed in containers whose vapour tightness is most often untested and the success is mainly trusted to low temperatures. Ultradrying has been very rarely used, perhaps by no more than 50 banks – mostly of wild species in botanical gardens—out of more than 1500 existing in the world. The UPM's results have proved beyond doubt the efficiency of ultradrying in orthodox seeds at least under anaerobic conditions and also prove the comparatively lesser importance of temperature for ultradried seeds.

It should be noted that moisture content of 4–5% may be enough for the efficient conservation of some orthodox seeds such as those of many legumes (Ellis & al. 1988). However, as this author points out, these seeds do not suffer when ultradesiccated to 1–3%. In a genebank, with thousands of accessions being handled, the need to individualise the methods should be avoided. In this light, a general benefit to the use of ultradrying can be attributed–always speaking of orthodox seeds.

ULTRADRYING WITH SILICA GEL

Ultradrying could be achieved following different ways, but perhaps the most practical one might consist of storing the seeds in balance with dehydrated silica gel. Leaving some gel afterwards with the stored seeds inside the tight container is most useful because the coloured indicator would show any accidental entry of moisture.

Advantages of the use of Silica Gel:

- It provides a practical method to desiccate the seed samples.
- It reaches moisture levels of 2-3%, lower than those obtained by most.
- Tkeeps these levels indefinitely in tight containers.
- It warns over possible anomalies (*i.e.* moisture intake) in the container—when used in combination with a coloured indicator.
- It can be regenerated after its use through a new dehydration process with heat.
- It delays also ageing by absorbing toxic gases produced during the process.

Summary of the advantages of using silica gel with an indicator. Besides those related to ultradrying itself, it is worth noting the last one (number 6), which refers to a totally independent protection mechanism. In seed samples stored within a closed container, damaging gases are generated during the aging process and as they accumulate, they prove deathly for the entire sample. The mortality, which might have been supposed a priori probabilistic for every individual seed, becomes accelerated at the end. Being capable of absorbing these gases, silica gel significantly delays aging and postpones the moment of death. Ethylene is an important component of this mixture of gases (Lee, & al. 2001).

Silica gel is simply silicic anhydride, SiO_2, though amorphous (non crystalline) obtained through an industrial process. It is granular in texture, white and very porous. It is this last characteristic which gives its absorbent properties. When it is well dehydrated and placed on a scale, it can be observed to absorb up to 20% of its own weight of water. If placed in a closed container such as a Kilner jar, it balances itself with the confined atmosphere until this reaches approximately 10–12% of relative humidity.

The name "gel" is here improperly used. We understand by gel some substance of a colloidal nature, which is not this case. However, the use of this term for this product has long been used and is now strongly established.

The colour usually shown by silica gel is due to an indicator, added to see directly when it is dehydrated and when it has absorbed moisture. For many years, cobalt chloride (Cl_2CO) has been used. This substance gives the dehydrated gel a strong blue colour and a pale pink colour to the gel having absorbed moisture. Recently, the European Union banned its use because of considering it carcinogen through inhalation. A search for new alternatives led to some iron salts, where the change in colour can be poorly distinguished. At present, the most advisable alternative is methyl violet, which gives the dehydrated gel an orange colour and a green colour to the hydrated gel.

The gel is manufactured in granules of different sizes. This size is not irrelevant because, to be used in Kilner jars, a gel with a size > 4 mm can be more convenient, while a 2-3 mm size can be better for glass vials. By hydrating with water coarse silica gel granules (be careful, heat is produced!) the grains get fragmented, and it is possible to screen it in fractions with smaller granule sizes.

Hydrated silica gel can be regenerated through heating. When it is accompanied with cobalt or iron salts, this is done at approximately 220 °C, thus a standard cooking oven can be enough. The process may last one or two hours - depending on the mass to be regenerated. With an organic indicator (*e.g.* methyl violet) a temperature of 110–120 °C must not been surpassed and the process will require comparatively more time.

FREEZE DRYING

Freeze Drying ("liophilisation") has been successfully used to obtain ultradry levels of moisture. This procedure consists in freezing the seed water and sublimating the ice afterwards. This technique was developed in the National Centre of Natural History (Paris) in the 1980s for long term storage of pollen grains (Schoenke & Bey, 1981. The Conservatoire Botanique National Méditerranéen (CBNM) started using it for seeds in Porquerolles in 1987.

It is assured that freeze dried seeds do not suffer any decrease in their viability because of the process. However, it does not seem to be the case for every species and the details when applying this technique must be optimized for every species (Anonymous, 2006). As it will be explained, freeze drying may have the advantage of achieving a direct ultradrying instead of one done in stages – as is generally the case with silica gel which needs to be regenerated every time it is used. In comparison, Freeze drying needs a rather expensive machine (Hu, X. & al. 1998). In turn, silica gel reaches lower water contents. Also, considering that silica gel is going to play an essential role afterwards in keeping a low moisture and monitoring the samples, it may be more practical a direct drying with gel. In other words, while freeze drying only ultradries, silica gel not only ultradries but also provides an efficient monitoring during subsequent storage.

THE LOW TEMPERATURE FACTOR

The use of low temperatures has been widespread in long term seed preservation. Standards set by the Food and Agriculture Organization of the United Nations (FAO) and the International Plant Genetic Resources Institute (IPGRI, now Biodiversity) advise a storage temperature of -18 0C or lower (FAO/IPGRI, 1994) and almost everybody now planning a genebank instinctively thinks of a cold room capable of at least reaching -20°C. Localisation of the new installations of the Nordic Seed Bank in a region of "permafrost" responds to the same idea.

Unfortunately, emphasis on low temperatures and neglecting low moisture (either because of not drying enough or because of the use of inappropriate containers) has brought that a significant number of the actual genebanks are worried with the burdensome need to rejuvenate their prematurely aged material.

Seeds which had been freeze dried in the Conservatoire Botanique National Mediterranéen (CBNM) of Porquerolles and then kept at room temperature for 10 years showed a slightly better behaviour than those kept under low temperature. In a much longer term, ultradry seeds with silica gel kept for 40 years at room temperature inside a closet (Pérez-García & al., 2007) did not behave very differently from those kept in the cold room. It appears hence clear that the main role attributed to low temperature during decades should now be replaced by a much more thorough attention to the moisture factor.

However, it is obvious that low temperature helps, and it would be irresponsible not to take it into consideration. In the last mentioned work, it is suggested that temperatures between –5 °C and +5 °C could very well be enough for orthodox ultradry seeds, thus saving considerable amounts of energy.

INOCULUM PREPARATION AND METHOD OF INOCULATION

The Preparation of the inoculum to be used in microbiological challenge testing is an important component of the overall protocol. Typically, for vegetative cells, 18–24 h cultures revived from refrigerated broth cultures or slants or from cultures frozen in glycerol are used. The challenge cultures should be grown in media and under conditions suitable for optimal growth of the specific challenge culture. In some studies, specific challenge organisms may be adapted to certain conditions. Such adaptation will be tailored to the specific food. For example, *E. Coli* O157:H7 May be acid adapted with the appropriate acidulant prior to use in the challenge studies on acidic products. Bacterial spore suspensions may be stored in water under refrigeration or frozen in glycerol. Spore suspensions should be diluted in sterile water and heat-shocked immediately prior to inoculation. Spores of *C. botulinum* should be washed thoroughly prior to use to ensure that no free botulinal toxin is carried over into the product undergoing challenge testing, and, if possible, the spores should be heat-shocked in the food to be studied. Quantitative counts on the challenge suspensions may be conducted to aid in calculating the dilutions necessary to achieve the target inoculum in the challenge product. Appropriate procedures and containment facilities should be used when carrying out challenge tests with certain pathogens.TheMethod of inoculation is another extremelyimportant consideration when conducting amicrobiological challenge study.

Every effort must be made not to change the critical parameters of the product formulation undergoing challenge. There are a variety of inoculation methods that can be used depending upon the type of product being challenged. In aqueous liquid matrices such as sauces and gravies with high a_w (> 0.96), the challenge inoculum may be directly inoculated into the product with mixing, using a minimal amount of sterile water or buffer as a carrier. Use of a diluent adjusted to the approximate a_w of the product using the humectant present in the food minimizes the potential for erroneous results in intermediate a_w foods.

In studies where moisture level is one of the experimental variables, the inoculum may be suspended in the water or liquid used to adjust the moisture level of the formulation. For batch type inoculations, the inoculum may be added directly to the product in a mixing bowl or container. For individual package or pouch type applications, the inoculum may be aseptically injected using a sterile syringe through the package wall containing a rubber septum. In solid matrices with a_w > 0.96, such as cooked pasta or meat surfaces, an alternative to the syringe method may be the use of an atomizer. An atomizer sprays the inoculum, which is suspended in sterile water or buffer, into the ground product or onto the surface of product.

Spraying should be done in a containment hood or using other protective devices to avoid worker safety issues related to creation of pathogenic. In all these applications, the smallest amount of water or buffer practical for suspension of the inoculum should be used. Inoculum may also be transferred using a velvet pad, paint pad, or similar fibrous cloth provided the method is calibrated and reproducible levels of inoculum can be delivered with minimum moisture transfer. Preliminary analyses should be done to ensure that the a_w or moisture level of the formulation is not changed after inoculation.

Products or components with a_w <0.92 may be inoculated using the atomizer method with a minimal volume of carrier water or buffer. Again, the product should always be checked to ensure that the final product a_w or moisture level has not been changed. A short, post-inoculation drying period for some products may be needed prior to final packaging. Alternatively, they may be inoculated with challenge organisms that have been suspended in carrier water or buffer that has been added to sterile sand, flour, or a powdered form of the product (for example, dried pasta), and allowed to dry. Lyophilized culture may also be used for some applications. Inoculum viability and population levels should be determined in advance of the study. The dried inoculum preparation should be added aseptically to the test product and shaken or agitated thoroughly for even distribution of the inoculum.

Enough product should be inoculated so that a minimum of three replicates per sampling time is available throughout the challenge study. In some cases, such as in certain revalidation studies and for uninoculated control samples, fewer replicates may be used.

DURATION OF THE STUDY

It is Prudent to conduct the microbiological challenge study over, at least, the desired shelf life of the product. It is even more desirable to challenge the product for its entire desired shelf life plus a margin beyond the desired shelf life because it is important to determine what would happen if users would hold and consume the product beyond its intended shelf life. Some regulatory agencies require a minimum of data on shelf life plus at least one-third of the intended shelf life.

Another consideration impacting the duration of the challenge study is the temperature of product storage. Refrigerated products may be challenged for their entire shelf life under the target storage temperature, but under abuse temperatures they are typically held for shorter time.

In certain foodservice venues, it may be convenient for the food establishment to hold specific refrigerated products at room temperature for short periods of time. For example, some fast food operations may find it convenient to hold processed cheese slices at room temperature for up to 8 h. This allows the cheese to temper and melt faster when preparing food items such as hot sandwiches. However, pathogens may be present on the cheese slices due to cross-contamination through handling in the restaurant, and therefore challenge testing will be needed to provide evidence that this practice is safe.

If the restaurant would like to hold the cheese slices at room temperature for an 8 h shift, the duration of the challenge study should be at least 12 h. This challenge study is performed to ensure that the rapid growth of pathogens does not occur if the cheese slices are cross-contaminated in the restaurant through handling.

It is also desirable to test the product over and significantly beyond its entire shelf life because sublethal injury may occur in some products. This can lead to a long lag period, where it may not be possible to culture the inoculum, but over time, a small number of the injured cells recover and grow in the product. This rebound, or "Phoenix" phenomenon, has been observed in a number of products (Jay 1996). If the product is not tested for at least its entire shelf life, it is possible to miss the recovery and subsequent growth of the challenge organism late in its shelf life.

The Frequency of testing is governed by the duration of the microbiological challenge study. It is desirable to have a minimum of 5–7 data points over the shelf life in order to have a good indication of the inoculum behaviour. Typically, if the shelf life is measured in days, the frequency of testing should be at least daily, if not multiple times per day. If the shelf life is measured in weeks or months, the test frequency is typically no less than once per week.

All studies should start with "zero time" testing, that is, analysis of the product right after inoculation. For some types of products, it may be desirable to also allow an equilibration period for the inoculum to adapt to the product before testing. It may be desirable to test more frequently (for example, daily or multiple times per day) early in the challenge study (that is, for the first few days or week), and then reduce the frequency of testing to longer intervals.

FORMULATION FACTORS AND STORAGE CONDITIONS

When evaluating a formulation, it is important to understand the range of key factors that control its microbiological stability. Intrinsic factors such as pH, aw, or preservative level may be key to preventing the growth of pathogens or to preventing spoilage that would influence the safety of the product during its intended shelf life. It is, therefore, important to test each key variable singly and/or in combination in the formulation under worst-case conditions.

For example, if the target pH is 4.8 + 0.2 and the process capability is within that tolerance range, it is important to challenge the product on the high side of that range (that is, pH 5.0). Similarly, if sorbic acid is used at a level of 0.15 + 0.05%, the product should be challenged at the low concentration of 0.10%. This is recommended to ensure that the challenge study covers the process capability range for each critical factor in the formulation. Relevant intrinsic properties such as pH, aw, and salt level should be documented for each study for future comparison and reference.

Test samples should ideally be stored in the same packaging as intended for the commercial marketplace. If the commercial product is vacuum or MAP packaged, then the samples used in the microbiological challenge study should be packaged under the same conditions using the same packaging film.

The storage temperature used in the microbiological challenge study should include the typical temperature range at which the product is to be held and distributed. A refrigerated product that may be subject to temperature abuse should be challenged under representative abuse temperatures. Products that may encounter high humidity environments should also be challenged under those conditions (Notermans and others 1993). Some challenge studies may incorporate temperature cycling into their protocol. For example, the manufacturer may distribute a refrigerated product under well-controlled conditions for a portion of its shelf life, after which the product may be subjected to elevated temperatures immediately prior to and during use.

SAMPLE ANALYSIS

Typically, in a Microbiological challenge study, the levels of live challenge microorganisms are enumerated at each sampling point. Usually, it is desirable to have at least duplicate and, preferably, triplicate samples for analysis at each time point. In cases where higher levels of certainty are needed, a larger number of replicates should be used or the study should be replicated. The selection of enumeration media and method (for example, direct plating versus Most Probable Number) is dependent on the type of pathogens or surrogates used in the study.

If the product does not have a substantial background *microflora,* non-selective media for direct enumeration may be used. In cases where toxin-producing organisms are used (for example, *Staphylococcus aureus* or *C. botulinum*), appropriate toxin testing should be performed at each time point using the most current validated method. Toxin levels may not always be tested at each time point in the study, but should be done at frequent enough intervals throughout the desired shelf life of the product to determine if that shelf life is acceptable. Where appropriate, resucitation methods may be used to avoid erroneous result.

It is Prudent to analyse the product, including uninoculated control samples, at each or selected sampling points in the study to see how the background microflora is behaving over product shelf life. For example, if a product has a high background microflora, it may suppress the growth of the challenge inoculum. In some cases, this is useful and desirable because the product spoils before pathogens can grow. In other situations, the background microorganisms may not be universally present, leading to a potentially false sense of security. Also, under some circumstances, the background microorganisms can change the formulation parameters in the product to favour or inhibit growth of the inoculum over time (for example, molds can raise product pH; lactobacilli can decrease product pH).

It is also important to track pertinent physico-chemical parameters of the product over shelf life to see how they might change and influence the behaviour of the pathogen. Understanding how factors such as aw, moisture, salt level, pH, MAP gas concentrations, preservative levels, and other variables behave over product shelf life is key to understanding the microbiological stability of the product.

DATA INTERPRETATION

Once the microbiological challenge study is completed, the data should be analysed to see how the pathogens behaved over time.

Trend analysis and appropriate graphical plotting (that is, semi-log plots) of the data will show whether the challenge organisms died, remained stable, or increased in numbers over time. In the case of toxin-producing pathogens, no toxin should be detected over the designated challenge period. Combining the quantitative inoculum data for each time point with data on the background microflora and the relevant physico-chemical parameters gives a powerful and broad representation of the microbiological stability of the formulation under evaluation. Based on these data, a reasonable shelf life can be established or adjustments can be made to the formulation so that it is less susceptible to pathogen growth.

When using microbiological challenge testing, as part of a process validation protocol, analysis of the data will show whether the process is capable of delivering the required level of lethality (that is, conforms with the pre-determined performance standard). Based on this information, adjustments can be made to the process, if necessary, in order to meet the lethality requirements. The Data from microbiological challenge testing can be used in developing predictive microbiological models or in validating existing ones. Predictive models are computer-based programmes that simulate or predict how specific microorganisms will behave in a formulation under specific conditions (for example, pH, a_w, moisture, salt, and preservatives). Microbiological challenge tests are used both to generate these types of empirical models and to validate their applicability.

Overall, well-designed challenge studies can provide critical information on the microbiological safety and stability of a food formulation. They are also invaluable in validating key lethality or microbiological control points in a process. Challenge studies can be an invaluable aid in determining if a food product requires temperature control throughout its shelf life or if it can tolerate storage at room temperature for a portion or all of its shelf life.

PASS/FAIL CRITERIA

Selection of microorganisms to use in challenge testing and/or modeling depends on the knowledge gained through commercial experience and/or on epidemiological data that indicate that the food under consideration or similar foods may be hazardous due to pathogen growth. In addition, the intrinsic properties (for example, pH, water activity, and preservatives) and extrinsic properties (for example, atmosphere, temperature, and processing) should be considered. The significance of a population increase varies with the hazard characterization of each microorganism. For example, the growth of infectious pathogens should always be controlled, whereas most toxin production requires substantial growth before a hazard exists. In this case, growth of the toxigenic organism alone does not result in a health hazard, but toxin production will.

The following list identifies microorganisms that can be used in a microbiological challenge study along with the panel's recommendations and rationale for selection and assessment of tolerable growth. Toxigenic molds such as *Aspergillus, Penicillium,* and *Fusarium spp*. Challenge studies related to the need for time/temperature control for safety are not recommended because mold provides a visual clue to prevent consumption of the spoiled product.

Bacillus Cereus: The absence of toxin formation is the preferred criterion. However, since toxin measurement is difficult, a 3 log increase over inoculum levels would indicate the need for time/temperature control.

This growth limit determination is based on the following:

- Typical initial levels of B. cereus are low; therefore, 1000 CFU/g would be a conservative initial level based on the literature.
- Populations of > 106 CFU/g are needed to produce toxin at levels hazardous to health (FDA 2001).
- The emetic toxin of B. cereus is heat stable; therefore, no reduction is likely.

Other considerations:

- *Bacillus cereus* spores are relatively heat sensitive.
- Baking is likely to destroy low levels found in flour used in bread products (Kaur 1986); however, unusual high heat resistance has been reported in pumpkin pie (Wyatt and Guy 1981).
- The potential for survival of B. cereus should be evaluated for specific products.
- Rice and potatoes have a history of association with *B. cereus* foodborne illness and therefore products containing rice or potatoes should be evaluated for time/temperature control requirements.

Campylobacter spp: No challenge testing is recommended because other organisms such as Salmonella have similar routes of contamination, are less fastidious, and are easier to culture. Furthermore, its minimal growth temperature and water activity of 32 °C (90 °F) and 0.98, respectively, make Campylobacter spp. it an unlikely candidate for challenge studies.

Clostridium botulinum: The absence of toxin formation based on current methodology is the recommended requirement. Other considerations: *C. botulinum* is appropriate to consider for certain cooked products, particularly those packaged under anaerobic and micro-aerophilic conditions such as MAP products; and those with a history of associated illness, such as products under oil and baked potatoes.

Clostridium perfringens: *A 3 log increase is recommended based on the following facts*:

- Although other products may contain surviving spores, *Clostridium perfringens* is relevant mainly to meat and poultry products, including sauces and gravies. Most products subject to the Food Code requirements will be either raw or freshly cooked.
- Vegetative cells of *C. perfringens* are easily destroyed by cooking meat and poultry products, and spore levels are typically low due to demanding sporulation requirements. An initial population of 100 CFU/g was considered to be a conservative worst case by the panel. A population of >105 CFU/g is needed to result in illness; therefore, a 3 log increase would control the hazard.

Enterohemorrhagic E. Coli: If modeling programmes are used to predict the growth of the pathogen, time/temperature holding conditions should maintain enterohemorrhagic *E. Coli* in lag phase due to the infectious nature of the microorganism. However, if laboratory challenge studies are used, the inherent variability in quantitative methods necessitates the use of a progressive increase of < 1 log as indicative that growth is controlled.

Listeria Monocytogenes: Recent risk assessments (FDA/USDA 2001) indicate that low numbers of *L. Monocytogenes* present a low risk to public health. In recognition of this, some countries such as Canada and Germany have established a tolerance for low levels of this organism in certain ready-to-eat foods that will not support growth to high levels. However, a tolerance for *L. Monocytogenes* has not been established in the United States for these types of foods. It is also recognized that products that support the growth of the microorganism present an increased risk. A *L. Monocytogenes* level of 100 CFU/g at the time of consumption may provide an acceptable level of consumer protection (Ross and others 2000). However, data are insufficient to determine general worst-case initial levels. Overall, the panel concluded that a 1 log increase was an appropriate level of control for *L. monocytogenes*. This level accounts for variability in enumeration techniques and represents a view that growth of this organism to high levels represents a risk to public health that must be controlled.

Salmonella spp: Appropriately validated pathogen modeling programmes for growth can be used to verify that*Salmonella* spp. is maintained in the lag phase. Otherwise, population growth should be limited to < 1 log, following the same rationale as for enterohemorrhagic *E. Coli*.

Shigella spp: No challenge studies are recommended for Shigella spp. because it has the same potential source as *Salmonella spp*. and has more fastidious growth and survival requirements.

Staphylococcus Aureus: No detectable toxin should be formed under the time/temperature studies evaluated. As with *C. Botulinum*, current methodology should be used for toxin detection and specific toxin levels should be determined. In lieu of testing for toxin, limiting growth to <3 logs may be used. This limiting growth level is based on an initial population of 1000 CFU/g, and a minimum of 10^6 CFU/g to produce toxin.

Other Considerations: *Staphylococcus aureus* is appropriate to study in foods that receive extensive handling because of the human source of the microorganism. *Staphylococcus aureus* does not compete well with other microorganisms; therefore, it is not appropriate to consider in foods with high levels of other organisms, such as raw vegetables or properly fermented products.

VALIDATION OF PROCESSING PARAMETERS

Establishment of traditional thermal processes for foods has been based on two Main Factors:

1. Knowledge of the thermal inactivation kinetics of the most heat-resistant pathogen of concern for each specific food product; and
2. Determination of the nature of heat transfer properties of the food system.

The validity of a thermal process must be confirmed by an inoculated challenge test conducted on the product under actual plant conditions using surrogate microorganisms as biological indicators to mimic pathogens. Thus, the two factors described above, which are well established for thermal processes, should be used for establishing and validating scheduled new thermal processes based on thermal effect on microorganisms, such as microwave heating.

For other preservation processes not based on heat inactivation, key pathogens of concern and nonpathogenic surrogates need to be identified and their significance evaluated. Surrogate microorganisms should be selected from well-known nonpathogenic populations, should mimic the target pathogenic microorganism in growth habits, should not be susceptible to injury, and should not exhibit non-reversible inhibition (thermal or otherwise). Surrogate microorganisms should be genetically stable and exhibit uniform thermal and growth characteristics from batch to batch over several generations. The durability to food and processing parameters should be similar to that of the target organism. Population of surrogates should be constant and maintain stable thermal and growth characteristics from batch to batch. Enumeration of surrogates should be rapid and should utilize inexpensive detection systems that easily differentiate them from natural flora.

Genetic stability of surrogates is desirable to obtain reproducible results. It also is recommended that surrogates do not establish themselves as "spoilage" organisms on equipment or in the production area. The validation process should be designed so that the surrogate exhibits a predictable time-temperature process character profile that correlates to that of the target pathogen. Introduction of system modifications or variables, leading to inaccurate results should be avoided (for example, thermocouple probes changing heating rates, nutrients added to the product for surrogate growth altering viscosity, and so on).

PROCESSING TECHNOLOGIES

WATER ACTIVITY AND PH

The Manipulation of water activity and/or pH is the less complicated of technologies in terms of equipment, expense, and expert personnel needed. Although it may not reduce the microbial load *per se*, reducing water activity or pH may retard or impede microbial growth. (For a more extended description on how water activity and pH can be used as preservation technologies and for a list of the optimum range pH and water activity for various pathogens of concern. When changing these characteristics of foods with the intention of safely storing a food at room temperature, those minimum pH and water activity values should be taken as guidance. At different temperatures and for different foods, these ranges may vary. For example, as the temperature moves away from optimum, a higher minimum pH is generally observed.

TECHNOLOGIES BASED ON THERMAL EFFECTS

In Addition to microbial inactivation by conventional methods of heating, microwave and ohmic and inductive heating are also considered to be heat-based processes that can inactivate microorganisms by thermal effects. Microwave and radio frequency heating refers to the use of electromagnetic waves of certain frequencies to generate heat in a material through two mechanisms-dielectric and ionic.

Ohmic heating is defined as the process of passing electric currents through foods or other materials to heat them. Ohmic heating is distinguished from other electrical heating methods by the presence of electrodes contacting the food, frequency, and waveform. Inductive heating is a process wherein electric currents are induced within the food due to oscillating electromagnetic fields generated by electric coils. No data about microbial death kinetics under inductive heating have been published.

For any of these heat-based processes, the magnitude of time/temperature history and the location of the cold points will determine the effect on microorganisms. The effectiveness of these processes also depend on water activity and pH of the product. Although the shape of the inactivation curves is expected to be similar to those in conventional heating, the intricacies of each of the technologies, however, need special attention if this technology is used for microbial inactivation.

For instance, in microwave heating a number of factors influence the location of the cold points, such as the composition, shape, and size of the food, the microwave frequency, and the applicator design. The location of the coldest-point and time/temperature history can be predicted through simulation softwares, and it is expected that food processors may be able to use them in the future. For determining the kinetics and efficiency of inactivation of microorganisms for these technologies, surrogate/indicator microorganisms could be selected from those traditionally used in thermal processing studies.

HIGH PRESSURE PROCESSING

High pressure processing (HPP), also described as high hydrostatic pressure (HHP) or ultra high pressure (UHP) processing, subjects liquid and solid foods, with or without packaging, to pressures between 100 and 800 MPa. Process temperature during pressure treatment can be specified from below 32 °F (0 °C) to above 212 °F (100 °C). Commercial exposure times can range from a millisecond pulse to over 20 min. Chemical and microbiological changes in the food generally will be a function of the process temperature and treatment time. The various effects of high hydrostatic pressure on microorganisms can be grouped into cell-envelope-related effects, pressure-induced cellular changes, biochemical aspects, and effects on genetic mechanisms.

HPP acts instantaneously and uniformly throughout a mass of food independent of size, shape, and food composition. Compression will uniformly increase the temperature of foods approximately 5 °F (3 °C) per 100 MPa. Compression of foods may shift the pH of the food as a function of imposed pressure and must be determined for each food treatment process. Water activity and pH are among the critical process factors in the inactivation of microbes by HPP. An increase in food temperature above room temperature, and to a lesser extent, a decrease below room temperature increases the inactivation rate of microorganisms during HPP treatment. Temperatures in the range of 113–122 °F (45–50 °C) appear to increase the rate of inactivation of food pathogens and spoilage microbes. Temperatures ranging from 194–230 °F (90–110 °C) in conjunction with pressures of 500-700 MPa have been used to inactivate sporeforming bacteria such as *Clostridium botulinum*. Current pressure processes include batch and semi-continuous systems, but no commercial continuous HPP systems are operating.

The Critical process factors in HPP include pressure, time at pressure, time to achieve treatment pressure, decompression time, treatment temperature (including adiabatic heating), initial product temperature, vessel temperature distribution at pressure, product pH, product composition, product water activity, packaging material integrity, and concurrent processing aids. Interestingly, because HPP acts instantaneously and uniformly through a mass of food, package size, shape, and composition are not factors in process determination. High hydrostatic pressures can cause undesirable structural changes in structurally fragile foods such as strawberries or lettuce (for example, cell deformation and cell membrane damage). Food products that have been brought to market include raw oysters, fruit jellies and jams, fruit juices, pourable salad dressings, raw squid, rice cakes, foie gras, ham, and guacamole.

A Biphasic pressure inactivation curve is frequently encountered for both vegetative bacteria and endospores indicating the residence of a small pressure-resistant sub-population. Tailing phenomena should be investigated carefully in challenge studies. The use of pathogens rather than surrogates for highly infective pathogens may be advised.

The Elimination of spores from low-acid foods presents food-processing and food-safety challenges to the industry. It is well established that bacterial endospores are the most pressure-resistant life forms known. One of the most heat-resistant pathogens, and one of the most lethal to human beings, is *C. botulinum,* primarily types A, B, E, and F. As such, *C. botulinum* heads the list of most pressure-resistant and dangerous organisms faced by HPP. Spore suspensions of strains 17B and Cap 9B tolerated exposures of 30 min to 827 MPa and 167 °F (75 °C) (Larkin and Reddy 1999; personal communication; unreferenced). Because some types of spores of *C. botulinum* are capable of surviving even the most extreme pressures and temperatures of HPP, there is no absolute microbial indicator for sterility by HPP. Among the sporeformers of concern, Bacillus cereus has been the most studied because of its facultative anaerobic nature and very low rate of lethality.

Normally, Gram-positive vegetative bacteria are more resistant to environmental stresses, including pressure, than vegetative cells of gram-negative bacteria. Among the pathogenic non-sporeforming gram-positive bacteria, *Listeria monocytogenes* and *Staphyloccocus aureus* are the two most well-studied regarding the use of HPP processing. *Staphyloccocus aureus* appears to have a high resistance to pressure.

There appears to be a wide range of pressure sensitivity among the pathogenic gram-negative bacteria. Patterson and others (1995) have studied a clinical isolate of *Escherichia Coli* O157: H7 That possesses pressure resistance comparable to spores. Some strains of *Salmonella* spp. have demonstrated relatively high levels of pressure resistances. Given these pressure resistances and their importance in food safety, *E. coli* O157:H7 and *Salmonella* spp. are of key concern in the development of effective HPP food treatments. For vegetative bacteria, nonpathogenic *Listeria* innocua is a useful surrogate for the foodborne pathogen, L. monocytogenes. A nonpathogenic strain of Bacillus may be useful as a surrogate for HPP-resistant *E. coli* O157:H7 isolates

Commercial Implications

Current practical operating pressures for commercial HPP food treatment intensifiers and pressure vessels are in the order of 580 MPa (85,000 psi). If this pressure is specified, then the following process times may be considered as first estimates for initial process planning. It must be understood that actual process parameters must be developed from challenge test packs.

Experience with acid foods suggests that shelf-stable (commercially sterile) products, having a water activity close to 1.0, and pH values less than 4.0, can be preserved using a pressure of 580 MPa and a process hold time of 3 min. This treatment has been shown to inactivate 106 cfu/g of *E. Coli O157*: H7, *Listeria* spp., *Salmonella* spp., or *Staphylococcus* spp. in salsa and apple juice.

Acid foods with pH values between 4.0 and 4.5 can be made commercially sterile using a pressure of 580 MPa and a hold time of 15 min. Products would have an initial temperature of about 71.6 °F (22 °C). Shorter hold times are possible if the product is to be refrigerated. Actual hold-time values must be determined from challenge packs and storage studies perhaps twice the length of the intended shelf life of the product.

Low-acid products can be rendered free of pathogens or pasteurized by HPP ; however, satisfactory guidelines for hold times at 580 MPa for low-acid food pasteurization have not emerged. For example, the post-package pasteurization of vacuum-packed cured meat products to eliminate *Listeria* spp. represents a useful application of HPP. Ground beef can be pasteurized by HPP to eliminate *E. Coli O157:* H7, *Listeria* spp., *Salmonella* spp., or *Staphylococcus* spp. Much more work is required to develop a suggested hold time at 580 MPa due to the potential for tailing. Changes in product colour and appearance may limit the usefulness of HPP treatment pressures above 200 to 300 MPa.

PULSED ELECTRIC FIELDS

High intensity pulsed electric field (PEF) processing involves the application of pulses of high voltage (typically 20–80 kV/cm) to foods placed between two electrodes. PEF may be applied in the form of exponentially decaying, square wave, bipolar, or oscillatory pulses at ambient, sub-ambient, or slightly above ambient temperature for less than 1 s. Use of PEF can reduce energy usage compared to thermal processes, as less energy is converted into heat, which also reduces detrimental changes to the sensory and physical properties of the food.

To date, PEF has been applied mainly to extend the shelf life of foods. Application of PEF is restricted to food products that can withstand high electric fields, have low electrical conductivity, and do not contain or form bubbles. The particle size of the food in both static and flow treatment modes is a limitation. Also, due to the variations in PEF systems, a method to accurately measure treatment delivery is still needed.

Factors that affect the microbial inactivation with PEF are process factors (electric field intensity, pulse width, treatment time and temperature, and pulse waveshapes), microbial entity factors (type, concentration, and growth stage of microorganism) and media factors (pH, antimicrobials and ionic compounds, conductivity, and medium ionic strength).

Many researchers have studied the effects of pulsed electric fields in microbial inactivation; however, due to the numerous critical process factors and broad experimental conditions used, definite conclusions about specific pathogen reductions cannot be made. Research that provides conclusive data on the PEF inactivation of pathogens of concern is clearly needed. Castro and others (1993) reported a 5-log reduction in bacteria, yeast, and mold counts suspended in orange juice treated with PEF. Zhang and others (1995) achieved a 9-log reduction in *E. coli* suspended in simulated milk ultrafiltrate treated with PEF by applying a converged electric field strength of 70 KV/cm for a short treatment time of 160 s. This processing condition may be adequate for commercial food pasteurization that requires 6- to 7-log reduction cycles (Zhang and others 1995).

However, numerous critical process factors exist and carefully designed studies need to be performed to better understand how these factors affect populations of pathogens of concern. Currently, there is little information on the use of surrogate microorganisms as indicators of pathogenic bacteria when PEF is used as a processing method. Selection of surrogates will require the prior identification of the microorganism of concern in a specific food and PEF system. The selection of the appropriate surrogate(s) will depend on the type of food, microflora, and process conditions (that is, electric field intensity, number of pulses, treatment time, pulse wave), and should also follow the general guidelines listed in the validation section.

IRRADIATION

Lrradiation of food refers to the process by which food is exposed to enough radiation energy to cause ionization. Ionization can lead to the death of microorganisms due to genetic damage which prevents cellular replication. For the treatment of foods, FDA has approved the use of gamma rays from decaying isotopes of cobalt-60 or cesium 137, x-rays with a maximum energy of five million electron volts (MeV), and electrons with a maximum energy of 10 MeV. An electron volt is the amount of energy acquired by an electron when accelerated by one volt in a vacuum. X-rays are produced when high energy electrons strike a thin metal film. Lethality of irradiation depends on the target (microorganism), condition of the treated item, and environmental factors. Addition or removal of salt or water, time/temperature of the treatment, or oxygen presence are factors that will influence the antimicrobial effect of irradiation. Irradiation is considered an additive in the U.S. and as such, it needs to be approved by the FDA office of premarket approval for each new application and labeled.

Two terms have been used to define the extent of pathogen reduction with irradiation. Radiation pasteurization refers to the destruction of pathogenic, non-spore-forming foodborne bacteria. In radiation pasteurization, medium dose treatments (1 to 10 kGy) reduce microbial populations, including pathogens in foods. Elimination of pathogens on meat, seafood, and poultry by medium dose irradiation has been studied. Sterilization radiation is used for radiation processes that will render the food commercially sterile or for foods that are both sterile and shelf stable. In this last case, sterilization must ensure the elimination of the most resistant pathogen, endospores of *Clostridium botulinum*. In order to achieve this, higher doses (42–71 kGy depending on the product) than the ones currently permitted for foods (up to 10 kGy, except for spices) are needed. Only frozen meats consumed by NASA astronauts have been permitted by FDA to be sterilized through irradiation.They are, however, in the market in other countries.

Lonizing radiation is used as a means of extending the shelf life of produce (Diehl 1995; Thayer and others 1996). FDA has approved the use of ionizing radiation with a range dose 0.3–1 kGy for growth and maturation inhibition. Not much effort has been applied to the control of foodborne pathogens on fresh foods, mainly because most medium and high level doses are not appropriate for produce since they can cause sensory defects (visual, texture, and flavour) and/or accelerated senescence (Thomas 1986; Barkai-Golan 1992). Ionizing irradiation has recently been used to eliminate E.Coli O157: H7 from apple Juice, and E. Coli O157: H7 and Salmonellae from seed and sprouts. Doses in the range of <1 to 3 kGy have been shown to reduce or eliminate populations of foodborne pathogens, postharvest spoilage organisms, and other microorganisms on produce (Moy 1983; Urbain 1986; Farkas 1997). Strawberry shelf life can be extended with treatments in the range of 2 to 3 kGy (Sommer and Maxie 1966; Zegota 1988; Marcotte 1992; Diehl 1995). Research conducted since that time suggests that irradiation can be an important treatment to enhance safety of other types of produce.

FDA and the U.S. Department of Agriculture Food Safety and Inspection Service have also approved irradiation to control foodborne pathogens in raw poultry with a dose range of 1.5–3.0 kGy. Recently, the irradiation of raw refrigerated and frozen meat has been approved with maximum doses of 4.7 and 7 kGy, respectively. Radiation doses of 2.5 kGy in beef will result in 6 log reduction of *Campylobacter*, 5 log reduction of *E.Coli* O157:

H7, 3 Log reduction in Salmonellaspp., and 5 log reduction of *Staphylococcus* cells (CAST 1996). Although the potential for consumer infection by pathogens is decreased greatly and shelf life is extended by radiation pasteurization of meat and poultry, the room temperature storage of raw meat products would be highly discouraged.

Other products, such as shell eggs (up to 3 kGy), have recently been approved for irradiation for safety reasons. Shell eggs can be irradiated with the intention of significantly reducing populations of *Salmonella* spp. Reduction levels depend upon radiation dose, initial level of pathogen contamination, or other treatment-related conditions.

The effect of irradiation on microbial populations suggests that it could also be used to decrease pathogensin a product with the intention of allowing microbiologically safe storage at ambient temperature for a specific time. However, the foods currently allowed to be irradiated are very limited. One could envision that in the future a food such as pumpkin pie could be irradiated to allow for safe ambient temperature storage. As with other technologies, organoleptic changes in the food would need to be considered. More important, the effectiveness of the technology will need to be validated for the specific application.

OTHER TECHNOLOGIES

Some of the technologies present greater limitations or are at a development stage that requires extensive further scientific research before they can be commercially used. For instance, high voltage arc discharge (application of discharge voltages through an electrode gap below an aqueous medium) causes electrolysis and highly reactive chemicals. Although microorganisms are inactivated, improved designs need to be developed before consideration for use in food preservation. Likewise, oscillating magnetic fields have been explored for their potential to inactivate microorganisms; however, the results are inconsistent. Data on inactivation of food microorganisms by ultrasound (energy generated by sound waves of 20,000 or more vibrations per second) are scarce, and limitations include the inclusion of particulates and other interfering substances.

Ultraviolet (UV) light is a promising technique, especially in treating water and fruit juices. A 4-log bacterial reduction was obtained for a variety of microorganisms when 400 J/m^2 was applied. Apple cider inoculated with *E. coli* O157:H7 treated in that manner achieved a 5-log reduction (Worobo 2000). Critical factors include the transmissivity, the geometric configuration of the reactor, the power, wavelength, and physical arrangement of the UV source, the product flow profile, and radiation path length.

RECALCITRANT SEEDS

These seeds are characterised by their relatively high moisture content (15–25%) and by a clear intolerance to desiccation. (Berjak and Pammenter, 2004). Therefore, they cannot be preserved by desiccation.Recalcitrant seeds can often be recognised morphologically because of their large size and smooth tegument. However, other seeds with a more normal appearance could also be recalcitrant. In the Mediterranean area they are not very spread but they can be found in woody species, climactic or not *(Quercus, Castanea)* or in riverbank species (Populus, Salix).In the rainforest, the species with recalcitrant seeds are dominant. Ecologically, they follow a strategy of lasting as seedlings, better than as seeds. That means that their seeds avoid getting too dry in a humid environment, where the very process of drying would be more difficult. Their lifespan is generally short, between a few weeks and 2–3 years.

If we make an exception of those species growing in riverbanks, seed size seems to be a key factor to determine the orthodox or recalcitrant behaviour of a species. This is because mechanical damage produced during drying is more severe in large seeds. For example, the seeds of pumpkin *(Cucurbita sp.)* and of some Phaseolus species (*P. lunatus)* can be damaged through drying while the smaller watermelon seeds *(Citrullus moschatus)* or those from green beans (Phaseolus vulgaris) are orthodox. The internal composition could also set differences. When working with small seeds, it is possible to be fairly confident that they will be orthodox, whereas, with an increase in size, exploratory tests should be carried out.`Being orthodox or recalcitrant is not an absolute character (Berjak & Pammenter,1994), since "semi-recalcitrant" seeds with an intermediate behaviour exist. These accept a partial desiccation, if only to a certain level. In any case, they should not be expected to tolerate ultradrying.

To Preserve recalcitrant seeds, there is not much to be done, except to put them in their lower tolerance limit as far as moisture content and temperature are concerned. As they inhabit wet places, it is usual that they die prematurely due to fungal or bacterial attack. This means that a treatment with an appropriate phyto-therapeutic product could perhaps extend their life span to some extent. Perhaps the most practical method whenever possible could be the establishment of gene banks of seedlings, since these seedlings use to be quite long lived when placed in low lighting conditions. For long term effects, it would be necessary to resort to the cryopreservation of their embryos.

It is not difficult to establish a protocol to set apart orthodox from recalcitrant seeds–based on the first ones being tolerant to desiccation.

This could consist of the following steps:

- An initial germination test,
- The desiccation of an equivalent lot and
- A new germination test of the desiccated seeds to show whether they are still alive or not. However, it should always be kept in mind that low values in the germination tests do not necessarily mean that the sample is dead or dying. These values could be originated by dormancy. If so, an immediate treatment of the seed (either by scarification, immersion in gibberellic acid solution or any

other suitable procedure) should follow in order to overcome any possible dormancy. This is particularly important for wild species.

Before deciding whether a given seed is recalcitrant - especially in medium sized seeds - there is still a test to conduct, which may enable us to desiccate it and to preserve it in the long term. Some apparently recalcitrant seeds seem to tolerate drying and perhaps ultradrying when these are gradually carried out, without sudden changes. For these "pseudo-recalcitrant" seeds it is advisable a previous successive balance with atmospheres showing a decreasing relative humidity. If this occurs, the number of species which could be preserved in the long term by the silica gel method will considerably be enlarged.

CRYOPRESERVATION

It consists of storing the material at temperatures near that of liquid nitrogen (–196°C). Under these conditions all enzymatic processes are practically halted and it is thought that any type of biological material (in plants: meristems, calluses, embryos, pollen, seeds, somatic tissues, etc.) can thus be preserved for a potentially infinite period. Only cosmic rays could place a limit after a few millennia.

Orthodox seeds can easily be cryopreserved. However, doing this has not much meaning once the high efficiency of ultradrying has been demonstrated for them. Ultradrying is also much cheaper. It is with recalcitrant seeds that cryopreservation might play a role in the future.

For an efficient cryopreservation, it is fundamental to avoid the intracellular formation of ice crystals which are highly damaging for the cell internal structures. A rapid descent of temperature is a must in the process, and a previous desiccation or ultradesiccation of the material is highly helpful. Also cryoprotective treatments (osmotic agents and/or substances such as glycerol, proline, abscisic acid, etc.) are often used. Most commonly used methods today aim to obtain a transition between an aqueous solution and an amorphous "glassy" state by means of concentrated solutions of sucrose and some derivatives of glycerol.

In the case of recalcitrant seeds many difficulties appear with whole seeds (Engelmann, 1997) but these can be partly overcome by preserving the embryos only (Fu & al., 1993). Basically, the technique consists of an encapsulation of the embryo in calcium alginate followed by a desiccation in concentrated sucrose solutions and then a final immersion in liquid nitrogen.

ADDITIONAL COMMENTS

Ultradrying and storage with silica gel of orthodox seeds in hermetic containers, preferably in an anaerobic atmosphere, with only moderately low temperatures, provide a tested and reliable method to keep the seeds alive for many decades and - likely - for centuries.

Efficient long term seed preservation saves much genetic material, and also much time, labour, electricity and money. The now widespread resignation over the need to rejuvenate the seed material in cycles of 25-35 years (in fact middle term preservation), could give path to the quite different situation of long term cycles of one or more centuries where the tremendous inconveniences of rejuvenation become minimised. Also, much of the time devoted to too frequent and tedious germination tests - so often only valid to check passively how the stored seeds are ageing and dying - can be devoted to research on more useful and stimulating subjects, such as the initial quality of the seed accessions, the possible dormancy in some of them and the adequate method to remove it, the orthodox, recalcitrant or pseudo-recalcitrant behaviour of others and so on (Gómez-Campo, 2006b).

Index